To Rosamaria and Franco,
two very dear friends,
With love
Huguette
December 2000

VISUAL COLOR AND COLOR MIXTURE

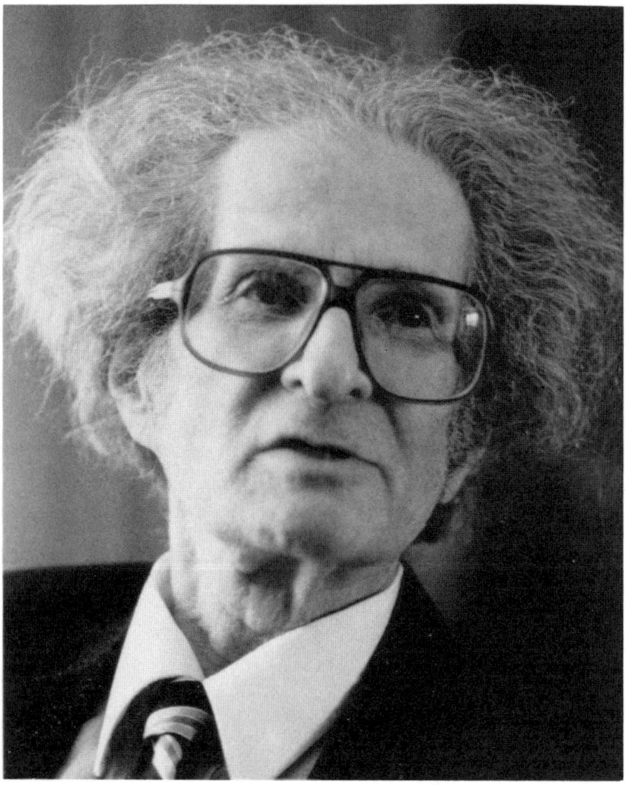

Jozef B. Cohen, 1921–1995 (Photo by Jerry Thompson)

VISUAL COLOR AND COLOR MIXTURE

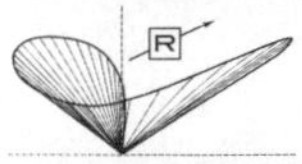

The Fundamental Color Space

Jozef B. Cohen

UNIVERSITY OF ILLINOIS PRESS
URBANA AND CHICAGO

Publication of this book was supported by a grant from the Eastman Kodak Company and by the University of Illinois Foundation.

Manufactured in the United States of America
♾ This book is printed on acid-free paper.

Library of Congress Cataloging-in-Publication Data
Cohen, Jozef, 1921–1995
Visual color and color mixture : the fundamental color space / Jozef B. Cohen.
p. cm.
Includes bibliographical references and index.
ISBN 0-252-02549-0 (hardcover : alk. paper)
1. Color. 2. Colorimetry. 3. Color vision. I. Title.
QC495.C73 2001
535.6–dc21 99-047681

C 5 4 3 2

To the great colorimetrist of the nineteenth century,
Hermann Günther Grassmann,

and to the great colorimetrist of the twentieth century,
Günter Wyszecki

A naturalist would scearce expect to see ye science of [colours] become mathematicall, & yet I dare affirm that there is as much certainty in it as any other part of Opticks.

—Isaac Newton to Henry Oldenburg,
editor of *Philosophical Transactions of the Royal Society (London)*, February 6, 1671/2

Contents

Acknowledgments

Jozef Cohen was a long-time friend whose scholarship I admired. When Jozef died in 1995, he left an unfinished book manuscript that was to have been the culmination of his work on color. Jozef's widow, Dr. Huguette Cohen, asked if I would oversee the completion of the work so that it could be published.

Not being an expert in color, color mixture, matrix **R**, or even being very knowledgeable concerning the historical context of Jozef's work, I set about finding people who would undertake the many tasks that remained. Fortunately, Jozef's work had progressed to the point that the manuscript was essentially complete; even so, considerable expertise was required to finish the work and respond to the comments of reviewers.

I am deeply indebted to Hugh Fairman (consultant, Princeton, New Jersey) who, through many conversations with Jozef in his last days, was able to fine-tune some of the prose in a way agreeable to Jozef. Fairman also exerted much effort in tracking down references and reorganizing some of the figures. Rolf Kuehni (DyStar L.P., Charlotte, North Carolina) wrote the Foreword and was tremendously helpful in his careful reading of the manuscript and in making many necessary changes. That effort was shared by Scott Burns and Edward Kuznetsov (University of Illinois at Urbana-Champaign). Rand Evans (East Carolina University, Greenville, North Carolina) wrote Jozef's biographical sketch, placing his work in a historical perspective.

Michael H. Brill (Sarnoff Corporation, Princeton, New Jersey) carefully reviewed the manuscript and wrote the Introduction, explaining the central themes of the book and its conceptual context.

The driving force for bringing the work to completion was Dr. Huguette Cohen. Her gentle insistence and persistence kept us all going, and her careful work on the References and in searching Jozef's files was invaluable. I am certain that had Jozef been able to foresee the crucial role she played in the completion of his work, Dr. Huguette Cohen would have joined Hermann Günther Grassmann and Günter Wyszecki to whom this book is dedicated.

MORTON W. WEIR
Urbana, Illinois, 1999

Jozef B. Cohen, 1921–1995

This book was to be Jozef Cohen's magnum opus. Unfortunately his death on August 18, 1995, did not allow him to see its publication and, in fact, prevented him from completing the book entirely. This work is certainly the culmination of a lifetime of scholarship and searching for answers to questions that appealed to him.

Jozef Cohen was born on July 21, 1921, in Brookline, Massachusetts. He attended school there and went off to do his undergraduate work in psychology at the University of Maine, before transferring to the University of Chicago. That university had been the center for and still bore traces of Chicago functionalism, the school of psychological thought that deals with the functions of consciousness—what the mind does for us. This was a fitting place for Joe to get his undergraduate degree since he had a practical view of things. He always had his eyes open for what the pure principles of science do for us.

Graduating from Chicago in 1942, Joe went to Cornell University for graduate work as a Susan Linn Sage Fellow. There he met Karl M. Dallenbach, who still represented the line of staunch experimentalism he had learned from Edward Bradford Titchener, founder of Cornell's department of psychology and whose purist structuralism had sought the exhaustive analysis of mental states for the pure joy of learning. At Cornell, Joe exercised his penchant for meticulous science. He continued to consider himself a psychophysicist and listed himself as such in *Who's Who in America* rather than as a psychologist. The term *psychophysicist* deals with the relationship between objects in the physical world and our consciousness of them. This is exactly what Joe's career dealt with in his experimental work.

It was at Cornell that Joe began researching in two areas that would lead him eventually to this book: constancy phenomena in vision and tristimulus representations of color mixtures. A paper from one of these areas, "A Nomograph for the Brunswik Ratio" (Cohen & Quinn, 1946), was concerned with the development of a nomograph for the calculation of the Brunswik ratio in practical situations. Another, "A Transformation Matrix of Color Mixture Data" (Cohen & Walker, 1946), shows a computation that will allow the determination of tristimulus values and trichromatic coefficients from color-matching data. Both studies show Cohen's work true to form: meticulous, mathematical, and practical. (A complete listing of Cohen's published work follows.)

One cannot credit either the University of Chicago or Cornell for Joe's penchant for accuracy and his intolerance for anything that was not exact in doing science or in his desire to find the real-world applications for his discoveries. But it was perhaps in these places that he found the names and concepts that set the direction for his proclivities.

After graduation from Cornell in 1945, Joe became a research staff psychologist in S. S. Steven's psychoacoustics laboratory at Harvard University. A year later he was called back to Cornell on the faculty. He stayed at Cornell until 1948, when he went to the University of Illinois. He stayed there the remainder of his career, becoming a full professor in 1969 and professor emeritus in 1991.

This book deals with color science, which was an outgrowth of Joe's interest in vision. But that was only one of the areas where he contributed. He was a pedagogue in the truest and best sense of that word. He sought to teach in both his classes and his publications. One of his important pedagogical contributions was his Eyewitness Series in Psychology, published by Rand McNally in the 1960s and 1970s. This was basically an introduction to psychology in ten separate monographs, aimed at the college population and the intelligent lay public. It was through these publications that I first became familiar with the work of Jozef Cohen. The span of his knowledge and the meticulousness of his language and concepts are evident in these works, just as they were in all of his other publications.

Joe contributed the article "Psychology" in the *Encyclopedia Britannica* in 1974 and so was the first source for many people around the world for the definition of the field and its contents.

Joe was also a historian. Certainly every scholar is a historian at heart, but Joe truly took it to heart and wrote widely in the history of ideas. That learning and perspective is evident in this book with the historical analyses he makes of the issues considered. He liked to take a historical approach to questions in order to place the current state of thought in proper perspective. Joe and I coedited the centennial volume of the *American Journal of Psychology* in 1987 (Cohen & Evans, 1987). I found in working with him just how careful Joe could be in historical matters and how voluminous his knowledge was in so many areas.

He returned to the questions of color science in his "Dependency of the Spectral Reflectance Curves of the Munsell Color Chips" (1964), which dealt with color constancies among reflected colors. For projected wavelengths, the trichromatic theory with its rather straightforward equations was adequate. Reflected colors, such as those produced by paint chips in the hardware store, were a much more complex matter. Joe used reflectance values from a sample of a large number of Munsell color chips that had been measured some years earlier. He selected 150 such values and ran a principal components analysis on them. The result was the finding that virtually all the cumulative variance was accounted for by the first three of the principal components extracted. It was all quite obvious, of course, after Joe had done it. The possibility of taking what appeared to be an almost insoluble pattern of complex relationships and reducing them to something fundamental and calculable fascinated him. In many respects, the work on spectral reflectance curves pointed Joe in the direction of matrix **R**.

The culmination of Joe's visual science work was that on matrix **R**. Since this book is about matrix **R**, I will not attempt to summarize it here. There are significant papers done with W. E. Kappauf that form the starting point of this book that should be acknowledged, however (Cohen & Kappauf, 1982, 1985a, 1985b), and also one by Joe (Cohen, 1988), and two others with Burns and Kuznetsov (Burns, Cohen, & Kuznetsov, 1989, 1990) that lay the fundamental framework for this volume. This book and the articles that preceded it all carry Joe through a familiar theme of the way he approached intellectual matters throughout his career. There is a search for fundamentals—an attempt to reduce the plethora of disparate facts into a few primary principles. All of this is placed in historical context, so the reader can see the continuous line between the beginnings of trichromatic theory and the present day. Finally, there is the extension of the theoretical and scientific work into the real world and its application into things as practical as color television. It is for this basic work that Jozef Cohen received the prestigious Macbeth Award from the Inter-Society Color Council in 1992 for developing the matrix-**R** method of colorimetric calculation.

Joe's work through much of his life went largely unappreciated, even maligned. He viewed the Macbeth Award as a vindication and a sign of acceptance. Nevertheless, the difficulty he had over the years gaining support for his work led him and his wife to create in 1995 the Jozef and Huguette Cohen Innovation Fund at the University of Illinois at Urbana-Champaign, aimed at helping to support researchers exploring ideas that are too innovative or too much outside their usual area of work to obtain grants from traditional sources.

Jozef Cohen was a perfectionist in his expectancies of himself and of others. It has been with trepidation that his friends and colleagues have undertaken the project of making this culminating work of his thought and labors available to the world. Joe would certainly have felt he could have done it better, and he doubtlessly could have, but this at least gets the broad strokes of his final work where it belongs: in the hands of scholars, scientists, and technologists.

We are all indebted to Huguette Cohen, Joe's wife of almost 40 years, for allowing this work to be done and being a motivator to ensure it is done right.

RAND B. EVANS
Greenville, North Carolina, 1999

Jozef Cohen's Published Work

J. B. Cohen. (1941). A scale for the measurement of attitude toward the aesthetic value. *Journal of Psychology, 12,* 75–79.

J. B. Cohen and L. Guttman. (1943). Multiple rectilinear prediction and the resolution into components. *Psychometrika, 8,* 169–183.

J. B. Cohen. (1944). Note on Carlson and Harrell's factor analysis of voting among congressmen. *Journal of Social Psychology, 20,* 313–314.

J. Cohen. (1946). Color adaptation to 1945. *Psychological Bulletin, 43,* 121–140.

J. Cohen. (1946). Color adaptation of the human eye. *American Journal of Psychology, 59,* 84–110.

J. Cohen & B. Quinn. (1946). A nomograph for the Brunswik ratio. *American Journal of Psychology, 59,* 146–149.

J. Cohen & R. J. Walker. (1946). A transformation matrix of color mixture data. *American Journal of Psychology, 59,* 299–305.

J. Cohen. (1946). History of a three-color mixer. *Science, 104,* 166.

J. Cohen. (1948). Color adaptation to stimuli of different spectral composition but equal tristimulus values. *Science, 108,* 159–160.

J. Cohen. (1948). Review of R. M. Evans. An Introduction to Color. *Science, 108,* 728–729.

J. Cohen & D. Gordon. (1949). The Prevost-Fechner-Benham subjective colors. *Psychological Bulletin, 46,* 97–136.

J. Cohen. (1949). Color vision and factor analysis. *Psychological Review, 56,* 224–233.

J. Cohen. (1949). An improved color mixer. *American Journal of Psychology, 62,* 418–420.

J. Cohen & D. P. Ogdon. (1949). Taste-blindness to phenyl-thio-carbamide and related compounds. *Psychological Bulletin, 46,* 490–498.

J. Cohen & D. P. Ogdon. (1949). Taste-blindness to phenyl-thio-carbamide as a function of saliva. *Science, 110,* 532–533.

J. Cohen & D. P. Ogdon. (1951). Concerning taste-blindness to PTC. *Psychological Bulletin, 48,* 419–420.

J. Cohen. (1954). Multichromatic colorimeters. *Illuminating Engineering, 49,* 50–58.

J. Cohen & W. A. Gibson. (1962). Vector model for color sensations. *Journal of the Optical Society of America, 52,* 692–697.

J. Cohen. (1963). Salivary conditioning in the XVIIth century. *American Psychologist, 18,* 69.

J. Cohen. (1964). Dependency of the spectral reflectance curves of the Munsell color chips. *Psychonomic Science, 1*, 369–370.

J. Cohen. (1967). Stimulus intensity and the visual generalization gradient to wavelength in the pigeon. *Psychonomic Science, 8*, 461–462.

J. Cohen. (1967). The synthetic-analytic character of color vision in the pigeon. *Psychonomic Science, 9*, 429–430.

J. Cohen. (1968). Review of G. Wyszecki and W. S. Stiles, *Color Science: Concepts and Methods, Quantitative Data and Formulas*. *American Journal of Psychology, 81*, 128–129.

J. Cohen. (1969). *Sensation and Perception I. Vision*. Eyewitness Series in Psychology, series separate 4. Chicago: Rand McNally.

J. Cohen. (1969). *Sensation and Perception II. Audition and the Minor Senses*. Eyewitness Series in Psychology, series separate 5. Chicago: Rand McNally.

J. Cohen. (1969). *Complex Learning (A)*. Eyewitness Series in Psychology, series separate 12A. Chicago: Rand McNally.

J. Cohen. (1969). *Complex Learning (B)*. Eyewitness Series in Psychology, series separate 12B. Chicago: Rand McNally.

J. Cohen. (1969). *Personality Dynamics*. Eyewitness Series in Psychology, series separate 17. Chicago: Rand McNally.

J. Cohen. (1969). *Personality Assessment*. Eyewitness Series in Psychology, series separate 18. Chicago: Rand McNally.

J. Cohen. (1970). *Secondary Motivation I. Personal Motives*. Eyewitness Series in Psychology, series separate 7. Chicago: Rand McNally.

J. Cohen. (1970). *Secondary Motivation II. Social Motives*. Eyewitness Series in Psychology, series separate 8. Chicago: Rand McNally.

J. Cohen. (1971). *Operant Behavior and Operant Conditioning*. Eyewitness Series in Psychology, series separate 11. Chicago: Rand McNally.

J. Cohen. (1971). *Thinking*. Eyewitness Series in Psychology, series separate 13. Chicago: Rand McNally.

J. Cohen & W. Little. (1971). Neural model for the Stevens' power law. *Perception and Psychophysics, 10*, 269–270.

J. Cohen. (1974). Psychology. *Encyclopedia Britannica* (15th ed.).

J. Cohen & T. P. Friden. (1975). The Euclidean nature of color space. *Bulletin of the Psychonomic Society, 5*, 159–161.

J. Cohen & T. P. Friden. (1976). Euclidean color space and its invariants. *Proceedings of the Technical Association of the Graphic Arts*, 411–429.

J. B. Cohen and W. E. Kappauf. (1982). Metameric color stimuli, fundamental metamers, and Wyszecki's metameric blacks. *American Journal of Psychology, 95*, 537–564. (Chinese translation in *Colour and Optics*, 1984.)

J. B. Cohen & W. E. Kappauf. (1985a). Color mixture and fundamental metamers: Theory, algebra, geometry, application. *American Journal of Psychology, 98*, 171–259.

J. B. Cohen & W. E. Kappauf. (1985b). Metameric color stimuli, fundamental metamers, and Wyszecki's metameric blacks. *American Journal of Psychology, 98*, 537–564.

J. Cohen. (1986). *Temas de Psicologia* (Vols. 1–10). Mexico: Editorial Trillas (Spanish translation of J. Cohen, *Eyewitness Series in Psychology*.)

J. B. Cohen & R. B. Evans (eds.). (1987). *100 Years of the American Journal of Psychology, Special Centennial Issue, 100*. Urbana: University of Illinois Press.

J. B. Cohen. (1988). Color and color mixture: Scalar and vector fundamentals. *Color Research and Application, 13*, 5–39.

S. A. Burns, J. B. Cohen, & E. N. Kuznetsov. (1989). Multiple metamers: Preserving color matches under diverse illuminants. *Color Research and Application, 14*, 16–22.

S. A. Burns, J. B. Cohen, & E. N. Kuznetsov. (1990). The Munsell Color System in fundamental color space. *Color Research and Application, 15*, 29–51.

Foreword

Color is one of the fundamental experiences of many living beings, from bees to humans. It is a means for those that have locomotion to be aware of certain reflectance structures of objects and materials in their line of sight. The mechanism of color vision has been of considerable interest to a number of history's great thinkers and scientists: Aristotle, Newton, Goethe, Young, Grassmann, Maxwell, Helmholtz, Schrödinger, and Hubel, to name a few.

The nature of the color experience plays a role in considerations of the puzzling questions of consciousness. Seemingly, however, it can play only a small role since, for example, in some manner bees must experience the reflectance structures of importance to them without the support of a cerebellum. There may be many refinements in human color vision experience compared to that of bees, but they certainly build on the kinds of experiences that many much simpler organisms must have.

The particular mechanisms of our color vision sense provide for many ancillary experiences: colored lights and with it the colors of the sky in many varied situations, the color of flames and fireworks, and color television.

How color vision works has been the subject of intense investigation since the time of Newton. On an empirical basis it has found its solution in the color-matching functions. The details of the mechanisms the output of which results in the color vision have become more and more of a certainty, supported with neurophysiological data concerning the three types of cones in the retina. The overlapping response functions of the cones make possible the extraction of a considerable amount of detail in the reflectance structures of objects. Thus, many different experiences can be created from the ratios of three signals. But the three signals also impose considerable limitations on the resolution of structural differences in the reflection properties of materials—hence, the phenomenon of metamerism, by which objects that appear identical in color in reality can have widely varying reflectance functions (a phenomenon from which image reproduction industries profit mightily).

Newton had a crucial insight into the mechanism of color vision when he decided that the best way to represent spectral colors was in circular form rather than on a line. Although the implication of complementary colors eluded him, the model implied the existence of metamers. Maxwell strengthened the idea that geometry could be used as a model for a human sensory mechanism by showing that the appearance of a given desaturated colored light could be matched not only by the appropriate mixture of two primary lights but also by a mixture of the appropriate spectral light together with its complementary, thereby giving further theoretical support to the idea of metamerism. (The term *metamerism* was coined by the chemist Wilhelm Ostwald in 1919.) A solid mathematical foundation for Newton's and Maxwell's barocentric system was provided by Grassmann and later by Schrödinger. On the experimental side, von Helmholtz and Maxwell provided visual colorimeters that demonstrated the fundamental nature of metameric matches.

The practical effects of metamerism began to be experienced more widely in the nineteenth century with the development of synthetic dyes. It is likely that printers of wallpapers may have been aware of them earlier but managed to minimize their problems by making sure they used identical pigments from the relatively limited palette of those available. The invention of a growing number of synthetic dyes for the dyeing of cotton, silk, and wool resulted in metamerism's becoming an everyday problem for the dyer. But the commonality behind reflectances and spectral powers that are seen as metamers was not well understood until Günter Wyszecki proposed a hypothesis in 1953: metamers consist of visually active and inactive components, with the active components identical and the inactive ones variable in a set of metamers.

That metamers in some fashion reveal fundamental properties of the color vision system was an early conclusion, but how to unravel its secrets mathematically? It is Jozef Cohen's enduring contribution to have done so. His matrix **R** is a meta-image of sorts (perhaps not unlike an intermediary layer in a neural network) of the mechanism that extracts the visually active components from reflectance structures. The impact on color science of this contribution has not yet been fully realized. Cohen not only provides a mathematically succinct model of Wyszecki's hypothesis but in the process reformulates colorimetry in a logically consistent framework. The power of his mathematical model is demonstrated in the similarities it extracts out of the explicit color vision models of past and present (see chapter 16), an analysis that awaits a champion to develop it more fully from the beginnings that Cohen provided toward the end of his life.

We are fortunate that Cohen worked diligently to leave his understanding of metamerism and colorimetry behind in a fashion that is mathematically impeccable and places all the developments fully and clearly into their historical context. In the process we can follow Cohen's fascination with the glories and foibles of the history of color science.

This book provides a rich lode of thought on color vision and colorimetry. Much of it awaits full exploitation, in particular Cohen's ideas on improvements in color signal transmission and color reproduction. It is a text that is warmly recommended not only to all colorimetrists and color scientists interested in greater transparency and progress in their field but also in particular to the members of the International Commission on Illumination (CIE) who must ponder the state of and progress with their system of standards.

ROLF G. KUEHNI
Charlotte, North Carolina, 1999

Introduction

This book was Jozef Cohen's last work before he passed away on August 18, 1995. In it, he sets out to reformulate the algebra of color matching and color mixing. The book is intended to inform practitioners in the field of colorimetry, experimental psychologists, and students of matrix algebra who want to see specific worked examples. It also comes at a time when computer solutions to problems in colorimetry add to the relevance of the digital methods implicit in the matrix algebra (starting from the expression of functions of wavelength as k-dimensional vectors). The mathematical sections require prior familiarity with matrix algebra, but undergraduates with a recent semester of linear algebra will have no trouble understanding them. Another possible audience comprises aficionados of the history and philosophy of science. The historical notes are extensive and useful.

Readers may note the abrupt ending of the book at the completion of Chapter 16. This is because Cohen intended there to be two additional chapters. We have their titles ("The Fundamentals of Color Order Systems" and "Matrix **R** Operations and the Transmission of Colorimetric Information"), but no text had been written. We can only guess at the intended contents of those two chapters, but it is likely that some of the contents may be found in Burns, Cohen, and Kuznetsov (1989, 1990).

Central Themes

Central to the book, and to Cohen's career, is the matrix-**R** formulation of the algebra of color matching. Matrix **R** has had a pervasive impact on the color science community, evidenced in Cohen's own work, in many more publications that cite this work, and in an award for matrix **R** given to Cohen in 1992 by the Inter-Society Color Council.

Although the algebra of color matching has been well understood for more than a century, Cohen's picture of it has been particularly thought provoking and attractive to a fairly wide audience. Before matrix **R**, the matching of metameric lights was represented by mapping their spectral power distributions to one point in a three-dimensional space of tristimulus values (the precursors of perceived colors). But spectral curves and points in 3-space evoke very different pictures. Cohen's method merged these two pictures into one: matching spectral power distributions of lights are mapped to a single spectrum, called a fundamental metamer, which is a sum of the color-matching functions weighted by three coefficients (with appropriate units). Matrix **R**, computed from the color-matching functions, performs this mapping from one spectrum to another. Essentially matrix **R** finds the projected shadow of a spectrum from a full function space (approximated by k-dimensions) to a three-dimensional hyperplane in that space that is spanned by the color-matching functions. This new picture has become a *lingua franca* in applied color science, in large part because of Cohen's published papers. Later investigators (e.g., Trussell, 1991; Imai & Berns, 1998) used the picture of projection operators in k-space to frame optimization and reconstruction algorithms in color, so the idea has had significant practical influence.

Besides evoking thought among color scientists and technologists, the matrix-**R** formulation offered to mathematicians and physicists an example of a space with both affine and Euclidean interpretations. The mathematician W. W. Sawyer (1981) wrote a paper about this insight of Cohen. For mathematicians, Cohen's picture of color-matching space is like a Necker cube: there are two interpretations, and it seems that one cannot attend to both at once. (A similar duality between affine and Riemannian geometry was used earlier as a basis for color theory [von Schelling, 1956], but may have been too complicated for the wide appeal enjoyed by Cohen's formulation.)

Although matrix **R** is the central theme of this book, it is by no means the only source of value here. For one thing, Cohen deals at length with ontological questions such as the nature (and units) of a tristimulus value. He does not resolve all the issues, but he inspired additional thought on the subject (Brill, 1996). Also, one cannot overstate the book's contribution to the history of color science as it interacts with related fields, offered in an informal and engaging way. I do not think there are similar compendia.

Conceptual Context

Visual Color and Color Mixture is prone to being misunderstood if it is taken out of historical context. It seems useful to lay to rest three potential points of criticism.

The first is that the algebra of quotient spaces unites the tristimulus and fundamental metamer pictures more compactly and elegantly than matrix **R**. In the quotient-space algebra (Suppes, Krantz, Luce, & Tversky, 1989), if the vector space of radiometric functions is denoted V and the subspace of residual spectra (orthogonal to the color-matching functions) is denoted W, then the subspace of fundamental metamers has codimension 3, and the quotient space V/W has dimension 3. The algebra of quotient spaces is indeed elegant and complete, but it is not a commonly used language among color scientists and engineers. Also, because of its abstractness, it is further from algorithmic implementation than matrix-**R** formalism. On the question of which formalism is more appealing to the engineer, history has voted de facto in favor of matrix **R**.

It may be worthwhile to note the formal point that Cohen's residual spectra (differences of metameric radiometric functions) do not exhaust the functions that are orthogonal to the color-matching functions. Because any radiometric function is everywhere nonnegative, and its fundamental metamer is zero at the ends of the visible spectrum, Cohen's residual must be nonnegative at the ends of the visible spectrum. An alternative construct that does not have this property is the linear combination of the first four daylight principal components that is orthogonal to the color-matching functions (Brill, 1987).

The second potential criticism is that the projection operator acts in a wavelength space with a Euclidean metric and refers this metric to color-matching space without acknowledging its arbitrariness. This statement is exactly true. To paraphrase, matrix **R** depends on whether, in discrete computations, one samples at equal intervals in wavelength or in a monotonic function of wavelength (such as wavenumber). But one has to remember that Cohen was the first to apply the projection operator formalism to color matching, and his contribution was the basis of a significant later advance (Fairman, 1987) that has no such arbitrariness. Fairman generalized matrix **R** to a matrix **S** defined as follows,

$$\mathbf{S} = \mathbf{B}(\mathbf{A}'\mathbf{B})^{-1}\mathbf{A}', \tag{0.1}$$

where $k \times 3$-matrix **B** replaces two of the instances of $k \times 3$-matrix **A** in the expression for matrix **R**, and $'$ denotes matrix transposition. There are two distinct interpretations of matrix **S** in a color-matching context:

1. Let **A** be the color matching functions (as in Cohen's theory) and **B** be the spectra of the primaries of a three-primary additive color system (such as a television monitor). If **N** is a radiometric function, the spectral power distribution $\mathbf{N}^* = \mathbf{SN}$ is the linear combination of television primaries (columns of matrix **B**) that match **N** as seen with color-matching functions defining matrix **A**. This was the interpretation discussed by Fairman, with consultation from Eugene Allen (as noted in Fairman's arti-

cle). Here, the selection of wavelength domain affects **B**, **N**, **N***, and **S**, but not **A**. It should be noted that the wavelength-domain selection appears consistently (*covariantly*) in **N*** given its appearance in **N** and **B**.

2. Let **A** be the color matching functions multiplied by an illuminant spectral power distribution, and **B** be three principal component reflectance spectra obtained from natural scenes. If **N** is an arbitrary spectral reflectance, the reflectance spectrum **N*** = **SN** is the linear combination of principal components (columns of matrix **B**) that match **N** as seen with the illuminant and color-matching functions defining matrix **A**. This interpretation could be useful for basis function theories of color constancy, particularly those that do not necessitate basis function expansions of the illuminant (see Brill & West, 1986). Here, the selection of wavelength domain affects **A** and **S** but not **B**, **N**, or **N***. It should be noted that **N*** is now invariant to the wavelength-domain selection.

The final potential criticism is that although the only empirical inputs to matrix **R** are from color matching, there are hints in the book that matrix-**R** formalism yields the quantities of color sensation. The title of Chapter 16 ("A Synthesis of Color Theories by Matrix-**R** Operations") certainly contains such a hint. However, there is also clear evidence (as in Chapter 2) that Cohen was careful in distinguishing the formalism of color matching from theories of color sensation. Furthermore, all the color theories examined in Chapter 16 are acknowledged to be equivalent through matrix-**R** operations, which is consistent with the fact that these theories share the same color-matching functions. How can one reconcile the matching-sensation dichotomy with the hint of a connection between them? Cohen's hints at the connection may be grounded in a belief that has not found rigorous support but is still defensible: insofar as the matrix-**R** (hence color-matching) formalism yields structures that are similar to certain components of color sensation theories (such as opponent colors), one has to allow the possibility that they are connected, if only by a loose ecological constraint.

These objections have been answered by referring to the past, present, and future of Cohen's picture: its past popularity (compared with more formal pictures), current refinement by others (to avoid arbitrary basis selections), and a possible future justification of the matching-sensation connection. By adopting such a historical perspective, most readers of the book will find a much richer understanding of the deep structures in colorimetry and of the opportunities still hidden therein.

MICHAEL H. BRILL
Princeton, New Jersey, 1999

VISUAL COLOR AND COLOR MIXTURE

1

Data, Notation, and Nomenclature

Following Isaac Newton, I have deliberately employed the vague term *sensorium* with reference to the inner locus of color sensations. The sensorium does not refer to the retinal cells, the optic nerve, or any specialized center within the brain. I am concerned only with the mathematical analysis of the color processing mechanism, not with its physiological description. Except where explicitly noted, I have employed the CIE 10° 1964 RGB and XYZ color-matching functions (CIE, 1986).

In 1966, when computers were novel and data were input with punched cards, Eugene Allen wrote a prescient article, "Matrix Algebra for Colorimetrists," published in the short-lived journal *Color Engineering*. He predicted that colorimetric applications were elegantly represented by matrix algebra, an unusual concept then, but now quite common and expanded to linear algebra and geometry. Allen predicted, "The new generation of computer-oriented colorimetrists will use matrix algebra as a daily tool." (For an early article concerning the application of matrix algebra to color reproduction, see Miller, 1941.)

I shall use a somewhat specialized mathematical notation and nomenclature so as to enhance understanding and forfend confusion. Matrices are represented by bold capital letters, as **A**, **E**, **F**, or **N**, often with a subscript, as $\mathbf{N}_B$. Column matrices are represented like other matrices, by bold capital letters, and not by lowercase italics, customary in some mathematical textbooks. The transpose of a matrix is represented by a prime, as $\mathbf{A}'$ and not $\mathbf{A}^T$ customary in several mathematical textbooks. The inverse of a matrix is the usual superscript, as $\mathbf{M}^{-1}$. Dimensionality is represented by lowercase italics, as k. The number of vectors represented is the number of columns in the matrix. An identity matrix is represented by a bold capital **I**. Matrix elements are represented by lowercase italics with subscripts; for example, the element in row 3 and column 1 in matrix **A** is a_{31}. Vectors are represented by capital letters with over-arrows, as $\vec{A}$. Vector lengths are represented by capital letters with over-arrows and with double side bars, as $|\vec{A}|$. The number of vectors represented is the number of columns in the matrix. From time to time, additional notation and nomenclature will be introduced.

I have adopted the radiometric unit as wavelength (in nanometers, nm), not as wave number (in frequency, Hertz, Hz—the reciprocal of wavelength). (See Wald, 1965.) This conforms to current de facto standards. Thus, the usual R,G,B ordering of red, green, and blue primaries, as currently practiced in colorimetry and the imaging sciences, has been reversed to B,G,R to conform with wavelength and not wave number. Stated differently, the blue primary will appear (say) in the 400-nm range of the abscissa and closer to the origin rather than the reverse.

The recommended CIE notation, adopted long before the prevalence of matrices and computers, is not completely consonant with modern mathematical notations. Confusion often exists between the names of color stimuli and their tristimulus values. In current CIE usage, capital italic letters may represent either of these very different entities. To avoid inherent perplexities, the names of all primaries and other color stimuli are symbolized by capital roman letters, as R,G,B, and X,Y,Z. Tristimulus values are represented by lowercase italic letters with subscript t, as x_t, y_t, z_t, and not the usual capital italic letters X, Y, Z. This is desirable, I believe, because a lowercase format for tristimulus values is commensurate with almost universal mathematical representation of coefficients in linear simultaneous equations; use of capital italics as coefficients presents an unfamiliar appearance. My use of lowercase italic letters for tristimulus values (with appropriate subscripts) does not conflict with the use of these same letters with overbars, as $\overline{X}$, $\overline{Y}$, $\overline{Z}$, to represent color-matching functions; tristimulus values and color-matching functions are the same numbers—tristimulus values by rows and the color-matching functions by columns. Tristimulus values should logically be represented by lowercase italic letters without subscripts, but these letters are commonly used for chromaticity coordinates on the CIE chromaticity diagram; chromaticity coordinates are completely foreign and meaningless in this work.

2

Contemporary Psychophysical Constructs in Colorimetry

Color Stimuli, the Physical Precursors of Color Sensations

A precise, unique description of any color stimulus comprises a radiometric function specifying irradiance at an arbitrary number of passbands spanning the visible spectrum, say, between 400 nm and 700 nm. As suggested, the passbands are equal wavelengths, as is common in color science. The number of passbands k is commonly about 31, but I have used as many as 300 for computations discussed in this book. Thus, every radiometric function requires k numbers referring to radiance. A radiometric function is represented by a column matrix **N**, often with a subscript, as $\mathbf{N}_B$. The number of passbands, or arbitrary divisions of the visible spectrum (k), equals the number of rows in matrix **N**.

Matrix **N** is designated monochromatic when a single entry is nonzero and all other entries are zero. Note that k entries are still required for complete specification—one nonzero value and also $(k - 1)$ zero values. Monochromatic stimuli with unit radiance as the nonzero value are commonly employed in color-matching experiments, as discussed in many prior articles, including Cohen (1988).

Matrix **N** is designated heterochromatic when more than a single entry is nonzero. Heterochromatic stimuli, devoid of nonzero entries, are said to be continuous. Hetrochromatic stimuli, with zero entries, are said to be discontinuous.

An **N** matrix signifying a radiometric function with positive or zero values is designated real, because in principle, this color stimulus may be realized as an actual physical stimulus. An **N** matrix signifying a radiometric function with at least one negative value is designated imaginary, because this color stimulus cannot be realized as an extant physical stimulus. Nevertheless, imaginary color stimuli often have theoretical importance in color science, as negative and imaginary numbers have theoretical importance in mathematics.

Examples of radiometric functions pervade the literature (see, e.g., Evans, 1948, 1974; Wyszecki & Stiles, 1982). Barnes (1939) assembled a splendid collection of reflectance functions describing artists' pigments, including cadmium orange, strontium lemon yellow, and cobalt violet. Nickerson (personal communication, 1955) recorded every reflectance function (radiometric functions when viewed under an equal-energy illuminant), in tabular form, from 380 nm to 770 nm, for every chip in the entire *Munsell Book of Color*. Early novel radiometric functions, where Fraunhofer lines specified wavelength, are found in Rood (1895). The first radiometric functions appear to have been recorded by John F. W. Herschel (1845).

The Stimulus Error

Colorimetrists are required to distinguish carefully between the physical entity of a color stimulus (radiant power) and the color sensation evoked by that stimulus. We are often predisposed to endow the objective physical stimulus with the imprint of the evoked subjective psychological sensation. This confusion between the properties of the stimulus and the properties of sensation was termed the *stimulus error* by E. B. Titchener (1905); the problem was classically deliberated by E. G. Boring (1921).

The stimulus error is exemplified by the biblical assertion that honey is sweet and that wormwood is bitter. In fact, these substances are neither sweet nor bitter. Honey and wormwood are physical chemical stimuli *evoking* subjective psychological taste sensations, including those described verbally by the words *sweet* or *bitter*. In color science, the stimulus error is especially bewildering when we speak effortlessly (say) of red raspberries and blue grapes. We mean, of course, that light entering the visual system, after modification by the physical reflectance of raspberries or grapes, evokes sensations described respectively by the words *red* and *blue*. There is no redness in raspberries, and there is no blueness in grapes. Circa 500 B.C., the Greek philosopher Democritus (Diels, 1934–1937; Cohen & Drabkin, 1948; Freeman, 1966) addressed the stimulus error in a surviving fragment on perception: "By convention there is sweetness, by convention bitterness, by convention color, but in reality there are only atoms and space."

Isaac Newton was well aware of the stimulus error in his 1704 *Opticks* (p. 90):

> If at any time I speak of Light and Rays as coloured or endued [*sic*] with Colours, I would be understood to speak not philosophically and properly, but grosly [*sic*], and according to such conceptions as vulgar People in seeing all these Experiments would be apt to frame. For the rays to speak properly are not coloured. In them there is nothing else than a certain power and disposition to stir up a sensation of this or that Colour. . . . Colours in the Object are nothing but a disposition to reflect this or that sort of rays more copiously than the rest; in the rays they are nothing but a disposition to propagate this or that Motion into the Sensorium, and in the Sensorium they are Sensations of those Motions under the forms of Colours.

Note Newton's emphasis of the word *sensation*, a critical and obvious construct to be rediscovered by Young, Helmholtz, and Maxwell.

Probably the most eloquent description of the colorimetric stimulus-sensation perplexity was formulated by Jean Le Rond d'Alembert, a scholar virtually unknown to colorimetric workers. In the eighteenth century, this revolutionary political theorist and mathematician wrote under the subject heading "Couleur" (d'Alembert 1751–1765) in that colossal compendium of knowledge known as Diderot's *Encyclopedia*:

> It is obvious that whiteness, or redness, and so forth, exists only within our minds and not at all in the [external] bodies to which they are attributed. [This is true] because of habits learned in our childhood. It is a very peculiar thing and worthy of the attention of metaphysics. We have a tendency to attribute to material substances that which really belongs to a mental state. And nothing is more extraordinary in the operation of our mind than to see it carry outside of itself and spread, if you will, its sensations upon a substance to which they [the sensations] cannot belong. (See also d'Alembert, 1761–1780, 1789-1797.)

The early nineteenth-century philosopher Arthur Schopenhauer, best remembered for his treatises advocating suffering and sacrifice to achieve death—the euthanasia of consciousness (the basis of Freud's concept of *thanatos* in contrast to Freud's *eros*)—traveled to Weimar to engage in personal conversations with Johann Wolfgang von Goethe. They spoke of color and color theory. Schopenhauer (1816) embraced the construct of the stimulus error in his semi-scientific essay, *Über das Sehen und die Farben*, as manifested in this excerpt: "Colors themselves, their interrelationships and the conformity of their appearance, all this resides in the eye itself, and is only a special modification of the activity of the retina. The external cause can only act as a stimulus [*sic*], the inducement to the manifestation of that activity; therefore, it [the stimulus] can act only in a subordinate way . . . similar to friction bringing about dormant electricity. . . . In no way, however, can colors . . . exist somewhere outside of the eye, purely objectively."

The Imprecise Nomenclature of Color Sensations

Colorimetrists contribute to the stimulus error because they often refer to both color stimuli and color sensations merely by name, as blue or green, or by a single letter, as B, G, X, or Y. These names or notations are confusing because they *may* designate the radiometric function of the stimulus (**N** matrices in this work), or they *may* designate the color sensation evoked by the color stimulus.

Color sensations have multitudes of names, and being totally subjective, we shall never determine whether the color sensation of one observer is the precise equivalent of the color sensation of another observer, even when both use the same descriptors.

Color names, of course, should be accompanied by some sort of color stimulus lest the name be totally without referent. (This problem has been admirably addressed by Birren, 1979.) The National Bureau of Standards (Judd & Kelly, 1939) correlated the color names of 13 separate color name collections to the Munsell renotated chips. The investigators found, for example, that a sensation described as "light, greenish yellow," referring to stimuli specified by Munsell renotation Y 8/5 to Y 10/8 viewed under daylight, is described variously by 92 other names, including chartreuse, liqueur, chelsea yellow, mellow yellow, and primrose yellow.

Color nomenclature is even more imprecise because of linguistic, sociological, and anthropological diversities. Apparently these deviations were first recognized by W. E. Gladstone (1858), Greek scholar extraordinary and prime minister to Queen Victoria, who noted peculiarities in Homer's use of the word *blue*. Subsequently these cultural color naming differences were recorded by Woodworth (1910), especially by Berlin and Kay (1969), Hickerson (1971), Bornstein (1973), Kay and McDaniel (1978), and Stanlaw (1987). To avoid this confusion, colorimetrists have developed techniques for specifying a color sensation (not a color stimulus) by the coefficients (tristimulus values) of a color-matching "equation" using three specified primary color stimuli. The CIE coefficients are the best-known tristimulus values and will be discussed. This book will introduce two new techniques for the mathematical specification of identical color sensations.

To its lasting credit, color science has avoided totally the necessity for describing color sensations by color names. It has adopted the so-called null method, whereby the observer simply reports when two sensations, evoked by two adjacent color stimuli, either match in *every* respect or do not match in every respect. If a discrepancy exists between the two sensations, the nature of the discrepancy is not communicated to the experimenter.

Metamerism and Color Mixture

Color science would be an elegant domain if a single color stimulus, specified by a unique radiometric function, evoked a unique color sensation. This, however, is contrary to our experience; an infinite number of color stimuli (described by different radiometric functions) evoke the *same* color sensation. These color stimuli were termed *metamers* by Wilhelm Ostwald (1919), the first Nobel laureate in chemistry. These color stimuli are *metameric* to each other, and they belong to the same *metameric suite.* Diverse methods for generating metamers have been reported by Morris (1947), Brewer and Holly (1948), Allen (1969), Takahama and Nayatani (1972), and Schmitt (1976, 1977a, 1977b). The most widely used technique was introduced by Manfred Richter (1958); later in this book, I demonstrate that Richter's method is a special case of my own matrix-**R** operations.

Figure 1 shows 3 metameric radiometric distributions, taken from a collection of 12 that were laboriously generated by Günter Wyszecki (1962) using Richter's method; all stimuli evoke the identical gray sensation when viewed under standard illuminant C. The three metameric stimuli are identified by identical tristimulus values: $x_t = 29.41$, $y_t = 30.00$, and $z_t = 35.43$.

Several years later, Wyszecki proposed that each of these metamers has a common fundamental, processed by the visual system, and a different residual, ignored by the visual system. Matrix-**R** operations mimic the color processing mechanism and isolate the common fundamental shown in Figure 1D and the respectively different residuals shown in Figure 1E–G. The fundamental plus the residual, wavelength for wavelength, equals the radiometric function; for example, the fundamental in Figure 1D plus the residual in Figure 1E equals the metameric function in Figure 1A.

The standard color-matching experiment, described in Cohen (1988), provides the simplest cases of metameric stimuli. Allowing negative coefficients, three monochromatic stimuli are compounded to evoke a sensation that matches the sensation of a single monochromatic stimulus. Here, two vastly different color stimuli evoke the same color sensation; therefore, the color stimuli are metamers. (Further discussion of the color-matching experiment appears later in this book.)

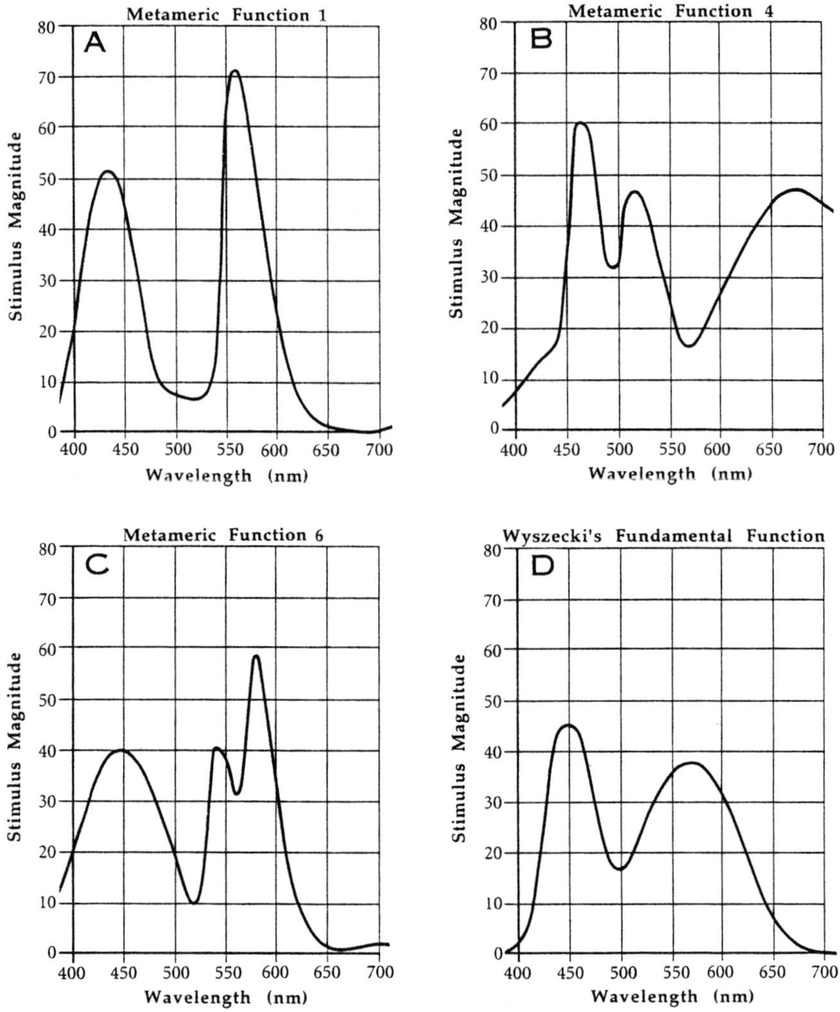

Figure 1. Matrix **R** applied to three metameric functions generated by Günter Wyszecki. *Continued next page.*

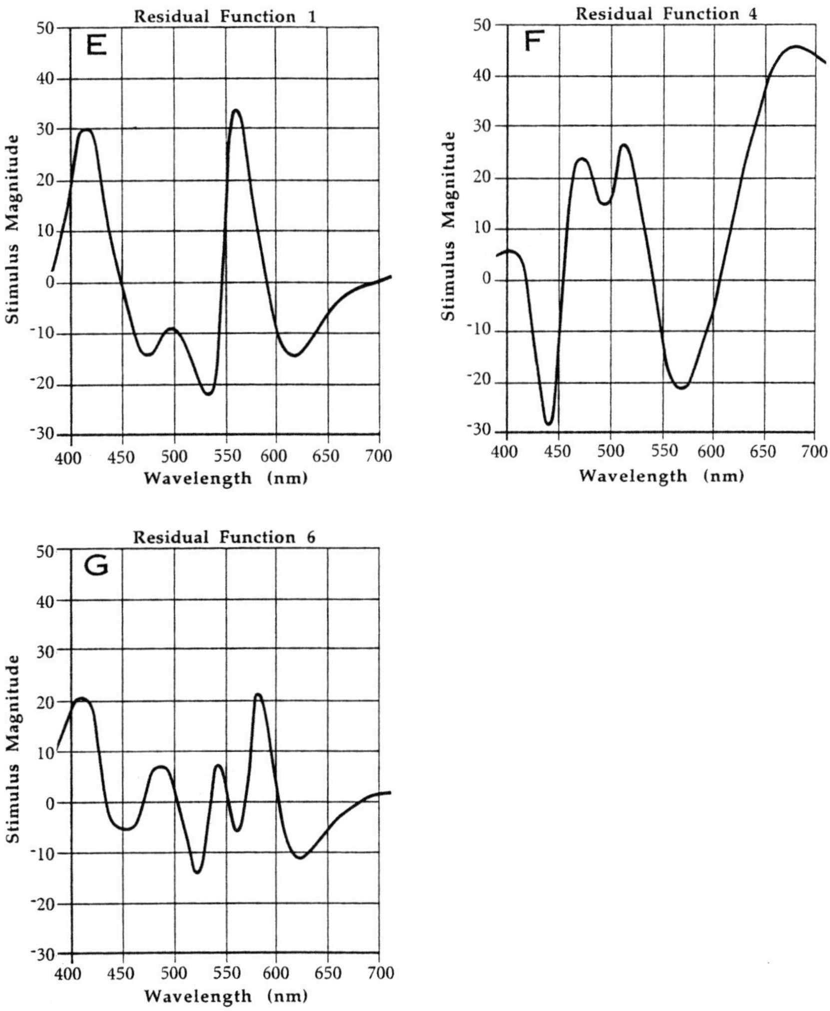

Figure 1. Continued.

3

The Newtonian Doctrine and Trichromacy

A meticulous formulation requires unqualified acceptance of two propositions (the *Newtonian Doctrine* in our nomenclature) established by Isaac Newton (1671–1672): (1) that the visible spectrum is continuous and infinitely divisible and (2) that to every spectral color stimulus belongs the same refrangibility and to every refrangibility belongs the same spectral color stimulus. Following empirical observations that multitudes of color sensations may be evoked by the additive compound of only three arbitrary color stimuli in varying magnitudes trichromacy demands that three "channels" or "dimensions" exist in the visual system. A hypothesis of more that three channels or dimensions conflicts with the construct of scientific simplification and economy, or, as usually stated, with the accepted tenet of Occam's razor—that we are not permitted more theoretical constructs than are absolutely necessary to explain the phenomena observed.

This book holds a strict definition of trichromacy: that trichromacy is the characteristic of the visual system that processes incoming color stimuli and is *not* a property of the stimuli themselves. This definition is critical to my development. Given the evidenced properties of the visual system, I shall demonstrate that other definitions cannot be entertained. By my analysis, trichromacy countenances no contradiction whatsoever; for example, it does not permit division of the physical spectrum into three primitive color stimuli or primaries.

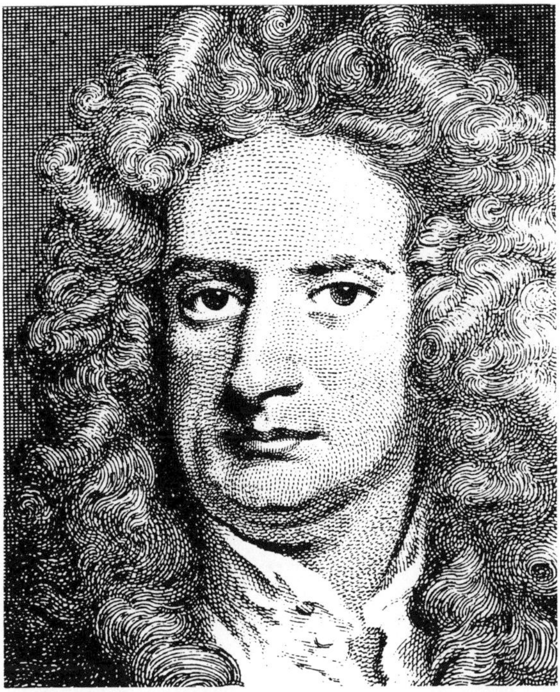

Sir Isaac Newton, 1642–1727

Weale (1957) and Lang (1980, 1983) have reviewed the historical aspects of trichromatic constructs in the seventeenth and eighteenth centuries; invariably researchers were severely influenced by the three-color mezzotint printing demonstrations of Jakob Christoffel LeBlon (1756). (See also Mortimer, 1731; de Montdorge, 1751; Singer, 1906; Burch, 1910; Peddie, 1917; Allen, 1969; Birren, 1981.) Early colorimetrists attributed trichromacy to the stimulus (Castel, 1723, 1735, 1737, 1740, 1743; Guyot, 1769–1770; Palmer, 1777, 1786; Marat, 1780, 1784; Wünsch, 1792; Lomonosov, 1747). (See also Schier, 1941; Walls, 1956; Leicester, 1970.)

Whence comes the magic number of three as the number of visual channels, in contrast to the Newtonian Doctrine that the visible spectrum was continuous and infinitely divisible? This is the legacy of Sir Thomas Young, a most honored naturalist of the late eighteenth and early nineteenth centuries. He is recognized generally for his definitive contributions to the undulatory theory of light, the first definitive anatomical description of the human eye, studies on blood and pus, tables used by insurance companies to assess life expectancy, navigation by

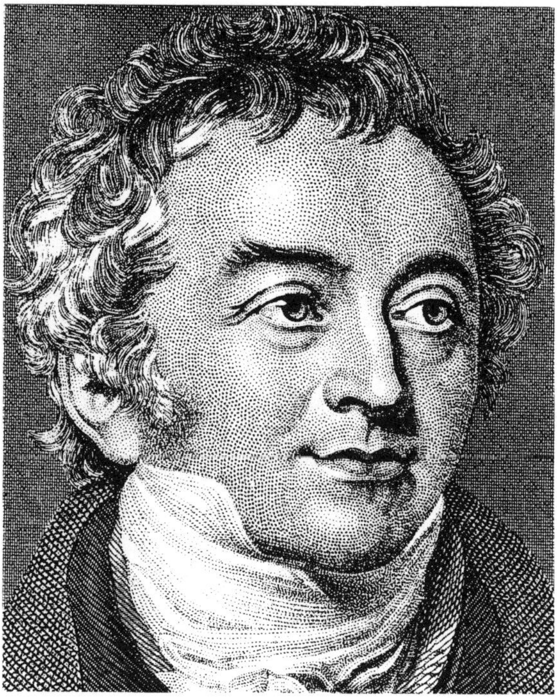

Sir Thomas Young, 1773–1829

latitude and longitude, and, in England, the translation of the Rosetta Stone (on the Continent, the credit is to Jean François Champollion). (George Peacock, 1855, and Alexander Wood, 1954, each wrote an extensive biography of Young. A number of articles document his testament: Arago, 1871; Mayer, 1875; Crew, 1930; Hecht, 1930; Parsons, 1930–1931; Millington, 1941.)

Young is usually credited with the first accurate statement of trichromacy embracing the Newtonian Doctrine. Both Maxwell and Helmholtz attributed the construct of trichromacy to Young. Helmholtz bestowed a magnificent posthumous honor by naming his own theory the *Young-Helmholtz*, as it appears even in elementary textbooks to this day, and not the *Helmholtz theory* or the *Helmholtz-Young theory*. Maxwell and Helmholtz were familiar with Young's writings only through Young's few publications.

This celebration of Young's contribution to color vision, viewed from the flat perspective of history, is interesting because Young misled himself and was inconsistent over time. His official interest in color vision began with the Bakerian Lecture to the Royal Society of London on November 12, 1801, published in 1802. Young (Arago, 1871), who held to the undulatory theory of light, supported his thesis by quotations from Newton, who held to the vibratory theory of light.

Young argued as an ancient scholar appealing to the authority of Aristotle. In this lecture, he gave his first interpretation to the mechanism of color as Hypothesis III: "The Sensations of different Colours depends on the different frequency of Vibrations, excited by light on the retina." Young quoted the 1672 article by Newton (1671–1672) in *Philosophical Transactions*:

> That fundamental supposition is, that the parts of the bodies, when briskly agitated, do excite vibrations in the ether, . . . which cause a sensation of light, by beating and dashing against the bottom of the eye, something after the manner that vibrations in the air cause a sensation of sound, by beating against the organs of hearing. Now, the most free and natural application of this hypothesis to the solution of the phenomena I take to be this: that the agitated parts of the bodies, according to their several sizes, . . . do excite vibrations in the ether of various depths or bignesses, which being promiscuously propagated through that medium to our eyes, effect in us a sensation of light of a white color; but if by any means those of unequal bigness be separated from one another, the largest beget a sensation of a red color, the least or shortest of a deep violet, and the intermediate one of intermediate colors.

Continuing this theme, Young transcribed a 1675 excerpt by Newton (1675–1676) included in Thomas Birch's *History of the Royal Society* and also an excerpt by Newton (1704) included in his *Opticks*.

Young then presented his theory of physiological trichromacy, precisely consonant with the definition I adopted. Here is the oft-quoted passage:

> Since, for the reason here assigned by Newton, it is probable that the motion of the retina is rather of a vibratory than an undulatory nature, the frequency of these vibrations must be dependent on the constitution of this substance. Now, as it is almost impossible to conceive of such a sensitive point on the retina to contain an infinite number of particles, each capable of vibrating in perfect unison with every possible undulation, it becomes necessary to suppose the number limited; for instance, to the three principal colors, red, yellow and blue.

Young held this strict conception of trichromacy for only a few months. William Hyde Wollaston (1802) delivered another address to the same Royal Society on June 24, 1802, and described his own experiments with prismatic colors; his researches, however, were markedly different from those of Newton. Wollaston rejected Newton's proposition concerning the continuity of the visible spectrum and its infinite divisibility (the Newtonian Doctrine). On the contrary, Wollaston reported four uniform and homogeneous spectral segments, described as demarcated by black lines, now known to be Fraunhofer lines (see Fraunhofer, 1888, 1898).

On repeating these prismatic experiments, Young (1802b) reported essential confirmation of Wollaston's observations. Young then rejected Newton's doctrine and, following Wollaston, concluded that the visible spectrum consisted of only three physical homogeneous segments, named red, green, and violet. These names derived from Young's new conception of a tripartite spectrum, and Young proposed corresponding retinal receptors; Young had abandoned the Newtonian Doctrine. This contrasts with Young's total acceptance of the Newtonian Doctrine before repeating Wollaston's imprecise experiments. Newton's Doctrine is correct, of course, and Young's position had become patently false.

In 1807, Young prepared *A Course of Lectures on Natural Philosophy and the Mechanical Arts* and once again gave his attention to the origin of color sensations. He omitted all reference to the retinal bases discussed in the Bakerian Lecture, but rather attended to Wollaston's experiments, clearly illustrated and showing three primary stimuli belonging to the spectrum. The text is illustrated with a color triangle and by disks with colored segments that, when rotated at high speed, fused to evoke a single compounded sensation. (The disks are usually called *Maxwell disks* because they were later used by James Clerk Maxwell.)

Here is a telling excerpt from Young's *Natural Philosophy*:

> It has generally been supposed since the time of Newton, that when the rays of light are separated as completely as possible by means of refraction, they exhibit seven varieties of colour. . . . The observations were, however, imperfect. . . . Dr. Wollaston has determined the division of the colored image or spectrum in a much more accurate manner than had been done before: by looking through a prism, at a narrow slit of light, he produces a more effectual separation of the colors than can be obtained by the common method of throwing the sun's image on a wall. The spectrum formed in this manner consists of four colors only, red, green, blue, and violet. . . . The colors differ scarcely at all in quality within their respective limits, but they vary in brightness. . . . A narrow line of yellow is generally visible at the limit of the red and the green, but its breadth scarcely exceeds that of the aperture by which the light is admitted, and Dr. Wollaston attributes it to the mixture of the red with the green light.
>
> Sir Isaac Newton observed that the effect of white light on the sense of sight might be imitated by a mixture of colors taken from differ-

> ent parts of the spectrum, not withstanding the omission of some of the rays naturally belonging to white light. . . . It is certain that the perfect sensations of yellow and of blue are produced respectively by mixture of red and green, and of green and violet light and there is reason to suspect that those sensations are always compounded of the separate sensations combined . . . and we may [therefore] *consider white light as a mixture of red, green, and violet only* [italics added].
>
> . . . The sensations of various kinds of light may also be combined in a still more satisfactory manner by painting the surface of a circle with different colours, in any way that may be desired, and causing it to revolve with such rapidity, that the whole may assume the appearance of a single tint, or a combination of tints. . . . From three simple sensations, with their combinations, we obtain seven primitive distinctions of colours; but the different proportions in which they may be combined afford a variety of tints beyond all calculation. The three simple sensations being red, green, and violet, the three binary combinations are yellow, consisting of red and green; crimson, of red and violet; and blue, of green and violet; and the seventh in order is white light composed of all three united.

Young continued to interpret trichromacy as a property of physical stimuli.

In 1817, Young's concept of trichromacy had matured, and he finally accepted Newton's doctrine. In "Chromatics," written as a supplement to the *Encyclopedia Britannica*, Young (1824) considered the trichromacy problem for the last time:

> There is little reason to doubt, that white light consists of an infinite number of rays, varying gradually among each other, without any marked distinctions, and continued on the one hand into the dark chemical rays, and on the other into the rays of invisible heat; and that all these varieties are separable from each other by refraction, and preserve always a distinct and constant refrangibility [the Newtonian Doctrine]. The species of homogeneous light, however, distinguishable from each other by the eye, are only five [and not seven]; red, yellow, green, blue, and violet. . . . The immediate effects of the combinations of the primitive colours on the sense of sight afford an illustration of some of the physiological [*sic*] characters of sensation in general. It is well-known that a mixture of red and green light produces a simple sensation, perfectly identical with that which belongs to the minute portion of yellow light originally found in the spectrum. . . . It is obvious, therefore, that the eye has no immediate power of analysing such light; and if we seek for the simplest arrangement, which would enable it to receive and discriminate the impressions of the different parts of the spectrum, we may sup-

> pose three distinct sensations [channels] only to be excited by the rays of the three principal pure colours, falling on any given point of the retina, the red, the green, and the violet, while the rays occupying the intermediate spaces are capable of producing mixed sensations, the yellow those which belong to the red and the green.

Nowhere during his lifetime did Young reference the precursors to trichromacy discussed above, but it is possible, and perhaps probable, that Young was familiar with a considerable portion of this prior work. Le Blon's process had been carefully demonstrated to the Royal Society and duly recorded in its official journal; Young had an opportunity to read about Le Blon in the same journal that published his own articles.

A more interesting possibility exists, however; I have assembled circumstantial evidence that Young was likely familiar with colored mezzotints depicting human anatomy, printed with three or four colored plates, by Jacques Gautier d'Agoty (1749a, 1749b, 1752, 1752–1755) in the library at Göttingen while Young was a medical student at the University. The library reports that Gautier d'Agoty's *Observations sur L'Histoire Naturelle* with anatomical three-color mezzotints (some surviving at Yale University) was part of the Göttingen collection during Young's residence, but the mezzotints disappeared at approximately the time of his departure; the library at Göttingen acquired another set some fifty years later.

My work descends directly from Young's last communication. I shall demonstrate an original mathematical interpretation of trichromacy using the modern mathematical projection operator, a relatively recent development in linear algebra. The problem may be stated thus: a color stimulus is specified by a radiometric function with k passbands, that is, k numbers or k dimensions are required for specification; the color processing mechanism, by the principle of trichromacy embodying the Newtonian Doctrine, comprises only three dimensions (channels). Therefore, the color processing mechanism cannot transmit all information of the k-dimensional stimulus toward the sensorium to evoke a color sensation. The color processing mechanism must select only portions of the color stimulus for transmission; the selection principles follow precise mathematical rules to maximize the physical colorimetric information transmitted, as I shall discuss.

Trichromacy cannot belong to the stimulus. If this interpretation were valid, metamers and metameric suites of color stimuli would be nonexistent, and metamerism would not be a property of color vision. Every pair of matching color stimuli would individually comprise precisely the same combination of homogeneous segments of the visible spectrum in identical magnitudes. Under these conditions, a given color sensation must be evoked by a single unique combination of spectral stimuli (a radiometric function). Under these conditions, one color sensation can be evoked by only one unique color stimulus; this, of course, is *not* the case.

Thus, we come to the crucial problem of modern colorimetry. The visual system has only three channels or dimensions, but it must process color stimuli with k dimensions. The visual system must select only portions of each color stimulus for transmission to the sensorium, and the rules for selection are described by modern mathematics. While the transmitted color information is maximized, the fundamental nature of the visual system ensures the existence of metamers; I shall discuss the mathematics of this process later in this work.

The number three has become the virtually uncontested hallmark of colorimetry. Over the centuries, color researchers have conjured concepts of triads, trinities, triplets, triples, cubes with three degrees of freedom, tridimensionalities, triangles, pyramids, and trinomials. As we know, three is the precise dimensionality of the extant universe, the most familiar of all Euclidean spaces. Researchers have theorized, from time to time, that this may not be coincidental, but that transmission of color information to the sensorium may depend on atomic changes in molecular structures where, as in the real world, the upper dimensionality is three.

4

The Electromagnetic Spectrum

The Newtonian experiment on the prismatic decomposition of white light into its "darker" components is arguably the most important single contribution to the scientific endeavor. From the Newtonian Doctrine emerged the extraordinary concept of a vast electromagnetic spectrum, absolutely essential in radio and television communication, where the visible spectrum is but a minute segment. From the Newtonian Doctrine emerged the line spectra of elements, and thence, through a stained glass window, the Bohr atom, as Jacob Bronowski (1973) has declared eloquently. Despite its critical importance to the advance of science, the Newtonian Doctrine was questioned and reexamined well into the nineteenth century. I speculate that the laws governing the vast radiometric spectrum were arrived at by examining the far smaller visual spectrum.

In England, the Royal Society of London had been chartered in 1662, and its official publication was the *Philosophical Transactions*. In France, the Académie Royale des Sciences was chartered in 1666, and its official publication was the *Journal des Sçavans*. Newton's article on the prismatic decomposition of light was the first major contribution to either of these journals and inaugurated a new phase in the march of science. Prior publications had been printed as books—as Copernicus's *De Revolutionibus* in 1543, Kepler's *Astronomia Nova* in 1609, Descartes's *Dioptrique* in 1637, and even Boyle's *Experiments and Considerations Touching Colours* in 1664. Now, however, each journal publication was widely distributed, and criticisms and debate appeared in succeeding issues of the same journal. Many philosophers availed this opportunity to pronounce their criticisms of the Newtonian Doctrine, and the critical sequence has been carefully recorded by I. Bernard Cohen (1958). In 1716, the Royal Society received a communication from Wilhelm Leibnitz, then editor of the journal *Acta Eruditorium*. Leibnitz forwarded the experimental results of one Edme Mariotte (1717), who had attempted replication of Newton's prismatic experiments; Mariotte was unable to verify the doctrine. Briefly, the Royal Society itself became doubtful, and accordingly, J. T. Desaguliers (1716) published a simple technique for substantiation in the *Transactions*. Desaguliers (1722) repeated Newton's optical experiments before the Royal Society on December 6 and 13, 1722, and demonstrated unqualified confirmation. (See also Desaguliers, 1734–1744.)

Nevertheless, the debate continued beyond the pages of the *Philosophical Transactions*. For example, Joseph Reade (1814) published an article with an extraordinary title: "Experiments Tending to Prove, That Neither Sir Isaac Newton, Herschel, Nor Any Other Person, Ever Decomposed Incident or Impingent Light into the Prismatic Colours." Once again, the spectrum was composed of only three primitive physical colors: blue, red, and yellow. The most vocal of these nineteenth-century critics was Sir David Brewster, who wrote on a vast array of subjects, including ventriloquism (1828). Brewster had considerable opposition, but one contribution (Brewster, 1831) was republished in *Transactions of the Royal Society of Edinburgh* in 1834, with the fascinating title, "On a New Analysis of Solar Light, Indicating Three Primary Colours, Forming Coincident Spectra of Equal Length." This article contains an unusual colored plate showing the complete visible spectrum with its full range of hues and then three other spectra—one depicting the red spectrum spanning the entire range of the visible spectrum, and similarly for the yellow spectrum and the blue spectrum. The magnitudes of the three primitive spectra are shown as three continuous functions bearing an astonishing resemblance to modern color-matching functions. Brewster was wrong, of course; he failed to discredit the Newtonian Doctrine. Simultaneously, however, additional evidence confirmed that the visible spectrum was continuous and infinitely divisible and, furthermore, was merely a small portion of the vast electromagnetic spectrum.

In 1800 (one year before Young's Bakerian address), William Herschel (1800a, 1800b, 1800c) investigated heat distribution in the differently refrangible colored monochromatic rays of the solar system. As shown in Figure 2, Herschel created a spectrum essentially as Newton described and placed the bulbs of several thermometers in different locations throughout the visible spectrum. Gradually he traced the distribution of heat in the solar spectrum. As the heat

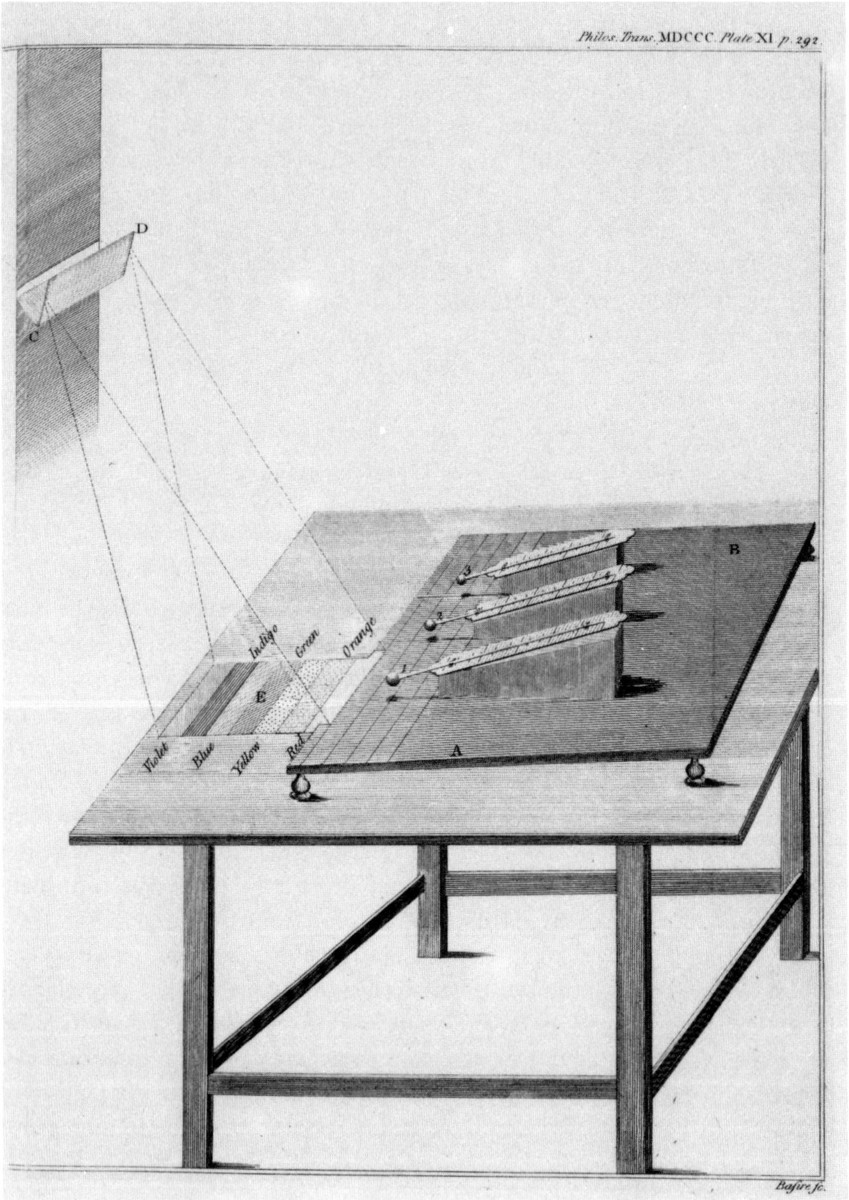

Figure 2. Engraving of Herschel's thermometers in the visible spectrum. (From Herschel, 1800b, 284–292.)

became more intense as he moved toward the red, the heat became even more intense as he passed beyond the red into that area of the electromagnetic spectrum now termed the infrared. He wrote, "It is evident that radiant heat is subject to the laws of refraction, . . . and perhaps lies a little beyond visible refraction. In this case, radiant heat will at least partly, if not chiefly, consist, if I may be permitted the expression, of invisible light." In this article, a paragraph is headed "Radiant Heat Is of Different Refrangibility." Herschel himself found no similar effect beyond the other end of visible spectrum, now termed the *ultraviolet*. (See Cornell, 1938a, 1938b.)

The 1807 volume of lectures by Thomas Young predicted portions of the continuous electromagnetic spectrum. In Young's lectures we read: "Dr. Herschel's experiments have shown that radiant heat consists of various parts which are differently refrangible, and that, in general, invisible heat is less refrangible than light. This discovery must be allowed to be one of the greatest to have been made since the days of Newton."

"It was first observed in Germany by Ritter, and soon afterwards in England by Dr. Wollaston, that the muriate of silver is blackened by invisible rays, which extend beyond the prismatic spectrum, on the violet side. It is therefore probable that those black or invisible rays, the violet, blue, green, and perhaps yellow, and the red rays of light, and the rays of invisible heat, constitute seven different degrees of the same scale, distinguished from each other into this limited number, not by natural division, but by their effect on our senses . . . " (See Strutt, 1889.)

In 1852, William Thompson, later Lord Kelvin, specifically extended the Newtonian Doctrine with respect to the undulatory theory of light: "It is assumed . . . that the undulatory theory of radiant light, according to which light is merely radiant heat, of which, the vibrations are performed in periods between certain limits of duration, is true. The chemical rays beyond the violet end of the spectrum consist of undulations shorter than those of the extreme visible light."

In the years 1865 to 1873, James Clerk Maxwell extended the undulations of the known continuous spectrum to the vast unknown of radio waves. Theoretically, he considered the effects of oscillating a current through a circular electric wire; today we call this device a *transmitter*. Under these conditions, electrical energy is lost and spreads about to form a fluctuating electromagnetic field. Theoretically, he considered the effects of the electromagnetic radiation on a similar circular electric wire at some distance; today we call this device a *receiver*. According to Maxwell, the receiver should oscillate with the same periodicity as the transmitter. This was the basis for the transmission and detection of all radiant energy.

In one way or another, all light is radiated by oscillators of high frequency, and the radiated energy must be detected by synchronous oscillators in the visual system. In theory, one could induce oscillations in an electric wire so that the radiations would evoke the sensation of, say, green, but available instruments cannot be made to oscillate at the required frequency. Therefore, light is radiated by some sort of atomic oscillators and detected by the visual system with mirror images of these same oscillators.

Heinrich Hertz (1894–1895), however, directly confirmed Maxwell's theory using a transmitter and receiver operating at frequencies far different from those of light. He used pulsed electrical charges with a frequency corresponding to a wavelength of about 60 cm. His corresponding receiver, placed at some distance, oscillated at precisely the same frequency.

The Newtonian Doctrine led to three kinds of spectra:

1. The continuous spectra, observed by Newton himself. The energy distribution of a spectrum, however, did not necessarily depend on the radiating substance, but rather on its temperature; this led to the concept of black-body radiators, extremely important in the development of experimental physics.
2. Line spectra, radiated by elements in a gaseous discharge tube and analyzed by a prismatic spectroscope, with a unique set of lines for each element. Niels Bohr explained these lines by the famous nuclear model bearing his name, a model that has become the logo of modern physics.
3. Solar spectra, which exhibit black lines, reported and explained by Joseph Fraunhofer. The lines represent the absorption of elements comprising the earth's atmosphere; a correspondence exists between the emission spectrum of an element from a discharge tube and its absorption of the same element in the atmosphere. (See Holton, 1952.)

All this from Newton's Doctrine! All this from Newton's interest in color! In the intervening years, scientists have exploited every aspect of these almost incredible derivatives of the Newtonian Doctrine, but we are still largely ignorant of the color processing mechanism itself.

5

The Wyszecki Hypothesis

Identical color sensations may be evoked by two metameric color stimuli described by different—sometimes vastly different—radiometric functions. In his 1953 doctoral thesis submitted to Manfred Richter, Wyszecki (1953) sought to explain the phenomenon of metamerism. His constructs, now termed the *Wyszecki hypothesis*, led directly to my matrix-**R** operations and the invariant, vector configuration of fundamental color space.

The Wyszecki hypothesis states that the visual color processing mechanism transmits only a portion of the color stimulus, while effectively rejecting or ignoring the remainder. The transmitted portion carries all information necessary for the color sensation and is termed the *fundamental.* The rejected or ignored portion carries no information whatever for any color sensation, but always evokes black—the null color "sensation," termed the *residual.* Wyszecki's original designators were "fundamental color stimulus" and "metameric black," respectively. One residual is trivial—the one that consists of all zeros. The fundamental itself is a radiometric function when the residual is trivial. In the years following Wyszecki's classical description of fundamentals, E. Hellmig (1954, 1955) contributed two articles on the residual, or metameric, black.

Note that the terms *metameric black* and *residual* are not quite synonymous. Metameric black is any radiometric function that does not evoke color sensation; residual is the metameric black component of a given radiometric

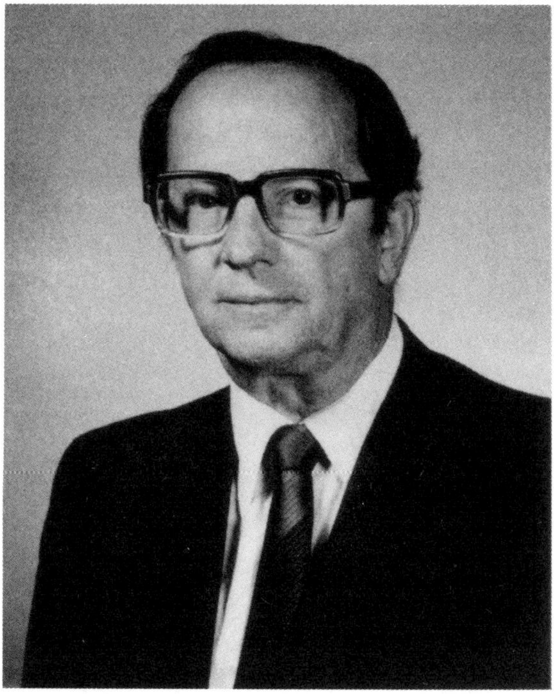

Günter Wyszecki, 1925–1985

function. However, except for cases where this distinction is important, the two terms are often used interchangeably.

Fundamentals may have positive and negative values; fundamentals of spectral stimuli always have both positive and negative values, and fundamentals of desaturated stimuli usually have all positive values. The fundamental of an equal-energy stimulus, for example, has all positive values. Residuals always have some negative values, and an ongoing debate still rages concerning the number and location of crossovers with zero radiation. (See Stiles & Wyszecki, 1968; Thornton, 1973, 1974, 1978, 1986a, 1986b, 1993a, 1993b; Ohta & Wyszecki, 1977; Kuehni, 1979; Brill, 1987; Berns & Kuehni, 1990; Kuehni & Berns, 1994.) The consensus suggests that residuals of real color stimuli almost always have 3 or 4 crossovers, while the residuals of artificial or imaginary color stimuli may have as many as 10 crossovers. Several artificial residuals are shown in Cohen (1988).

Clearly, two metameric color stimuli comprise the same fundamental but different residuals. Each member of a metameric suite has the same fundamental but a different residual.

The Wyszecki hypothesis may be stated in terms of matrices. Let **N** be the radiometric function specifying any color stimulus. Let matrix **N*** specify the fundamental of matrix **N** and matrix **B** specify the residual of matrix **N**. Then, by the Wyszecki hypothesis,

$$\mathbf{N} = \mathbf{N}^* + \mathbf{B}. \tag{5.1}$$

This equation states that every **N** matrix may be divided into two unique components: the fundamental, **N***, and the residual, **B**. The components add, element for element, to the radiometric function. Figure 1A–C illustrates three of Wyszecki's (1962) metameric radiometric functions. Figure 1D illustrates the fundamental common to all 12 of Wyszecki's metamers, including the three shown in Figure 1A–C. Figure 1E–G illustrates the residuals of the radiometric functions in Figure 1A–C; the resolutions were made by matrix-**R** operations.

The specification of a radiometric function requires k independent numbers. The fundamental also has k numbers, but only three are independent. The residual also has k numbers, but only $(k - 3)$ are independent. Thus, the independent values of the fundamental added to the independent values of the residual must equal the independent values of the radiometric function.

The fundamental, **N***, and the residual, **B**, themselves are forms of radiometric functions with the mathematical properties of radiometric functions. The addition (or subtraction) of one fundamental $\mathbf{N}_1^*$ and a second fundamental $\mathbf{N}_2^*$ gives $\mathbf{N}_3^*$, still another fundamental, the stimulus representing the additive color mixture of $\mathbf{N}_1^* + \mathbf{N}_2^*$. Similarly, the addition (or subtraction) of one residual $\mathbf{B}_1$ and a second residual $\mathbf{B}_2$ gives $\mathbf{B}_3$, still another residual. These operations are analogous to the addition or subtraction of radiometric functions.

Furthermore, a fundamental or residual may by multiplied (or divided) by a constant c to give another fundamental or residual, as a radiometric function may be multiplied (or divided) by a constant c to give another radiometric function. The operation produces a function proportionately larger or smaller than the given function.

Wyszecki was unable to resolve *any* **N** matrix into its components, **N*** and **B**, a procedure that awaited matrix-**R** operations.

Although Wyszecki did not isolate even a single fundamental, he did compute residuals for metamers generated by Richter's procedure. Given the radiometric functions of the metamers, Wyszecki was able to calculate additional members of the metameric suite. The subtraction of one metamer from another produced a residual, as explained by the following analysis.

There are two metameric color stimuli, $\mathbf{N}_1$ and $\mathbf{N}_2$, which have identical fundamentals. Then

$$\mathbf{N}_1 = \mathbf{N}_1^* + \mathbf{B}_1 \tag{5.2}$$

$$\mathbf{N}_2 = \mathbf{N}_2^* + \mathbf{B}_2 \tag{5.3}$$

$$\mathbf{N}_1 - \mathbf{N}_2 = \mathbf{N}_1^* - \mathbf{N}_2^* + \mathbf{B}_1 - \mathbf{B}_2 \tag{5.4}$$

$$\mathbf{N}_1 - \mathbf{N}_2 = 0 + \mathbf{B}_1 - \mathbf{B}_2 = \mathbf{B}_3 \tag{5.5}$$

But $\mathbf{N}_1^* = \mathbf{N}_2^*$ and in equation 5.4, $\mathbf{N}_1^* - \mathbf{N}_2^* = 0$, because the fundamentals of metameric stimuli are identical, and therefore, $\mathbf{N}_1 - \mathbf{N}_2$ is another residual. Wyszecki then added (or subtracted) these residuals to (or from) other members of the metameric suite (radiometric functions), and the added (or subtracted) residual fused with the incumbent residual, and a new metamer was created.

Has the reader noticed the slight misuse of terms in the discussion? From time to time I have referred to what is, in a strict sense, a member of the metameric suite as a metamer. The CIE (1987) defines metamers as spectrally different color stimuli that have the same tristimulus values; the corresponding property is called *metamerism*. Thus, at a minimum, a metamer consists of two radiometric spectra, and the concept of metamerism deals with the relationship between the two spectra, not to a property of any one spectrum. So a single radiometric spectrum, to "speake properly . . . and not grosly," as Newton put it, ought not to be referred to as a metamer. Colorists, however, are given to speaking grossly, and the term *metamer* is often substituted when a member of the metameric suite is meant. Just as is the case with making the stimulus error, it is acceptable to speak this way where the context does not demand differentiation between a member of the suite and the relationship between two members of the suite.

Note that adding residuals, to either a fundamental or a metamer, may produce imaginary radiometric functions with negative values. In that instance, a positive or negative constant may be used as a multiplier to modify the residual, so that the resulting metamer has all positive values.

We differentiate between scalar and vector quantities. In the simplest terms, a scalar quantity is characterized by magnitude only, and a vector quantity is characterized by both direction and magnitude. For example, mass and speed are scalar quantities; force and velocity are vector quantities.

Wyszecki's fundamentals may be described as either scalars or vectors, depending on their representation. Wyszecki conceived of fundamentals in the scalar mode. Cohen and Kappauf (1982, 1985) conceived of these same fundamentals in the vector mode where each color fundamental (and its sensation) is represented by a unique vector in an invariant, three-dimensional Euclidean vector configuration, termed *fundamental color space*. In the vector mode, the additive compounding of color fundamentals is described by vector addition, that is, color fundamentals (and color sensations) are added precisely as force vectors are compounded in mechanics. This explicit analogy to force vectors was proposed by James Clerk Maxwell (1860) more than a century ago.

The scalar representation of the fundamental has an interesting and critical representation in the vector mode relating to the scalar (or inner or dot) product, defined by Hermann Günther Grassmann (1847, 1853) in a prize-winning essay in 1833. The essay extended the geometric system sketched by Leibnitz in a letter to Huygens in 1679. The scalar product is the foundation of fundamental color space.

The scalar product $(\vec{V}_1 \cdot \vec{V}_2)$ is defined as

$$\vec{V}_1 \cdot \vec{V}_2 = |\vec{V}_1| \cdot |\vec{V}_2| \cdot \cos\phi, \tag{5.6}$$

where $\vec{V}_1$ and $\vec{V}_2$ designate any two vectors, $|\vec{V}_1|$ and $|\vec{V}_2|$ designate the lengths of the respective vectors, and ϕ is the intervening angle. The vectors are orthogonal (perpendicular, at 90 degrees and independent) when the scalar product is zero, except for the trivial case where one vector length is zero. (A detailed examination of the scalar product appears later in this book.)

Given any vector (signifying one color fundamental and one color sensation) in the vectorial fundamental color space, the scalar fundamental is the locus of scalar products between the given vector and each vector representing a spectral color fundamental (or sensation). This remarkable property is true of every vector in the configuration, real and imaginary, including the spectral vectors themselves.

Wyszecki's residuals also have two forms. The first is the scalar mode, as already described and as conceived by Wyszecki. The second is the vector mode where each residual (and its black sensation) is represented by a unique vector in an invariant $(k-3)$-dimensional Euclidean vector configuration, termed the *residual space*. In the vector mode, the additive compounding of residual stimuli is analogous to vector addition, that is, residual stimuli are added exactly as force vectors are compounded in mechanics.

The scalar residual has an analogous representation in the vector mode. Given any vector (signifying one residual and its black color sensation) in the residual vector space, the scalar residual is the locus of the scalar products between the given vector and each vector representing the residual component of each spectral color stimulus (or sensation). Again, this remarkable property is true of every vector in the configuration.

The vectors of fundamental color space lie in a three-dimensional configuration. Vector lengths and angular separations are preordained; this configuration is absolutely invariant and *independent of the primary triplet* selected for the color-matching experiment. An investigator cannot alter this configuration, in contrast to the arbitrary, investigator-dependent vectors and configurations of affine space as, for example, the CIE X,Y,Z space (x_t, y_t, z_t space, in my notation). Although the vectors and angles of fundamental color space are invariant, the basis or reference vectors must still be selected by the investigator; these may be any three independent vectors—either orthogonal or oblique. Their selection does not change the vector lengths or angles within the configuration. Again, the residual space has the same characteristics, except that $(k-3)$ basis vectors are required.

Wyszecki (1958, 1962) laboriously generated an entire collection of residuals from color-matching equations, and the residuals were carefully recorded for the use of colorimetrists. Now, every vector in fundamental color space is orthogonal to every vector in residual space. In higher dimensions, this relationship may be conceived, although not imagined. Presently I shall give pictorial representation to the three-dimensional case.

6

Color Matching

Spurious and Valid Color-Matching Equations

I begin with additional notation necessary for this section.

Matrix $\mathbf{N}_p$, $k \times 3$, is any independent triplet of column matrices comprising three radiometric functions, say, $\mathbf{N}_B$, $\mathbf{N}_G$, $\mathbf{N}_R$, representing three color stimuli used, for example, as primary color stimuli in a color-matching experiment.

Matrix **A**, $k \times 3$, emerges in a color-matching experiment as I have indicated, where ordered monochromatic standard stimuli (one for each band in the visible spectrum) are matched by three arbitrary primary stimuli, usually monochromatic. The rows of matrix **A** are weighting coefficients (percentages), or *tristimulus values*. The columns of matrix **A** are *color-matching functions* and are meaningful only as a triplet; a single color-matching function (one column of matrix **A**) has almost no significance by itself. The color-matching experiment begins with matrix $\mathbf{N}_p$ and ends with the entries in matrix **A**. Matrix **E** comprises the fundamentals of the primary radiometric functions in matrix $\mathbf{N}_p$, while matrix **A** is the matrix of color-matching functions emerging from a color-matching experiment. In the color-matching *equation*, the row entries (tristimulus values) of matrix **A** must weight the column entries (fundamentals) of matrix **E**. Matrix **A** must not be confused with matrix **E**.

Colorimetrists have a compelling and obsessive desire to write the rows of matrix **A** *alone* as a series of equations.

Here is an "equation" of lower class in CIE notation:

$$CC \equiv BB + GG + RR. \tag{6.1}$$

Here is the same "equation" in my notation:

$$c_t\mathrm{C} \equiv b_t\mathrm{B} + g_t\mathrm{G} + r_t\mathrm{R} \tag{6.2}$$

This spurious "equation," 6.1, is a recipe with no mathematical import. The capital Roman letters merely indicate names for the primaries, and the italicized letters represent the tristimulus values of a row in matrix **A**. Equation 6.2 states merely that c_t percentage of a color stimulus named C is matched by ($\equiv$) b_t percentage of a primary stimulus named B (blue), *added* to g_t percentage of a primary stimulus named G (green), *added* to r_t percentage of a primary named R (red). My notation is virtually universal in modern linear algebra but not in colorimetry. The quantities are not defined. "Equation" 6.1 is merely descriptive; it does not balance mathematically or physically. Nevertheless, it is repeatedly found in the literature, as, for example, in Wright (1958) and in Feynman, Leighton, and Sands (1963).

Here is an "equation" of a somewhat higher class:

$$1\mathbf{N}\mathrm{C} \equiv b_t\mathbf{N}_\mathrm{B} + g_t\mathbf{N}_\mathrm{G} + r_t\mathbf{N}_\mathrm{R}. \tag{6.3}$$

The names for the standard and the primary have been replaced by **N** matrices representing their radiometric functions. This equivalence is still not a valid mathematical equation; it does not balance mathematically or physically. The equation merely states that one color stimulus (specified by its radiometric function) is matched ($\equiv$) by the compounding of three other weighted stimuli (specified by their radiometric functions).

Finally, here is a valid color-matching equation of the highest class:

$$\mathbf{N}^*_\mathrm{C} = b_t\mathbf{N}^*_\mathrm{B} + g_t\mathbf{N}^*_\mathrm{G} + r_t\mathbf{N}^*_\mathrm{R}. \tag{6.4}$$

The radiometric functions of the standard and the primaries have been replaced, respectively, by the fundamentals of the standard and the primaries, as demanded by the Wyszecki hypothesis. Because the visual system processes only the fundamentals and is oblivious to the residuals, the standard and the primaries are effectively their respective fundamentals alone. This is a mathematically valid and legitimate equation. The equation states that the two sides of the bipartite field evoke identical color sensations *and* that the two sides are equal ($=$) mathematically and physically. This equation is acutely important in the development of matrix **R**.

Matrix **A**, with tristimulus values as rows, is described by an important characteristic colorimetric theorem. Given any **A** matrix, the algebraic sums of each column are the respective tristimulus values of an equal-energy stimulus. The tristimulus values of each monochromatic stimulus are the three entries in one row of matrix **A**, and because an equal-energy stimulus is the sum of all monochromatic stimuli spanning the visible spectrum, the tristimulus values of an equal-energy stimulus are exactly the sum of the tristimulus values for each primary, or, stated differently, the sum of each of the three columns in matrix **A**.

The CIE had a strong disposition to require that the mixture of "equal quantities" of the X,Y,Z stimuli combine to compound a stimulus that evokes the "white" color sensation. Therefore, the sums of the X,Y,Z columns were adjusted to be almost equal; the sums of columns could have been made exactly equal by multiplying each column with an appropriate constant; this would have no effect on the fundamental color space computed from matrix **A** by matrix-**R** operations. However, equal tristimulus values do not mean that "equal quantities" of the primaries are required to compound a stimulus that is equivalent to white sensation evoked by an equal-energy stimulus, as will be demonstrated. Indeed, very different magnitudes of X, Y, and Z are required.

This error is pervasive. For example, Wright (1958) in his *The Measurement of Colour* wrote:

> It has therefore become the practice to base the units of [primaries] (*R*), (*G*), and (*B*) on a match on a white of some defined quality. Since white may be regarded as a colour in which neither red nor green nor blue predominates, the arbitrary but reasonable convention that equal amounts of (R), (G), (B) are required to match white may be adopted. It should be noted that this is not a statement of the kind which future investigation might subsequently show to be slightly inaccurate. . . . It is merely a convention which has been adopted because it is convenient and because it fits with our conception of white as a neutral colour possessing no attribute of hue.

Wright continues that the amount of C is given by the sum of the tristimulus values. However, the tristimulus values are dimensionless percentages and cannot be added meaningfully. In equations 6.2, 6.3, and 6.4, one can add percentages of the same fundamental (or radiometric function, or residual) but not of different fundamentals (or radiometric functions, or residuals), as is the practice in computing the CIE chromaticity coordinates. Furthermore, the stimulus error is pervasive in Wright's argument. When the white sensation is devoid of blue, green, or red hues, one cannot conclude (from this observation alone) that white comprises *neither* blue, *nor* green, *nor* red components.

Characteristics of Color-Matching Equations

Equation 6.4 seems equivalent to equations in vast numbers of textbooks on linear algebra—well, not precisely, and I demonstrate with a simple example adapted from Strang (1988). Here are two simple simultaneous equations with an exact and unique solution:

$$\begin{aligned} 2x - y &= 1 \\ x + y &= 5. \end{aligned} \tag{6.5}$$

The solution is $x = 2$ and $y = 3$, as may be confirmed by insight or by the usual techniques, such as adding one equation to the other followed by back substitution. The equations may also be solved with geometry by plotting two lines in a two-dimensional cartesian frame, each line representing the locus of all solutions to one equation. Since a unique solution exists, the lines intersect at a single point, representing a solution to both equations simultaneously; the values x and y are the cartesian coordinates of the common point. This solution, however, is more a function of logic than of geometry. I plot all solutions to the first equation and all solutions to the second equation, and the solution to both is read from the common intersection point. We learn nothing of the internal vector structure describing these equations.

With respect to colorimetry, these two equations have a far more interesting format when written as a single vector equation:

$$\begin{bmatrix} 1 \\ 5 \end{bmatrix} = \begin{bmatrix} -2 & 1 \\ 1 & 1 \end{bmatrix} \begin{bmatrix} x \\ y \end{bmatrix}. \tag{6.6}$$

Each of these two equations has the same format as color-matching equation 6.4, and, as expected, the pair have the solution, $x = 2$ and $y = 3$. Each column is represented by a vector with appropriate coordinates in the two-dimensional cartesian frame illustrated in Figure 3. These are the initial functions, before multiplication by the tristimulus values, here called x and y.

The left column vector is analogous to the fundamental to be matched, and the two right column vectors are analogous to two primaries. The coefficients x and y are analogous to tristimulus values. In this example, we are given the primary fundamentals, and we seek to calculate the weighting coefficients or tristimulus values. Figure 3A shows that the vector sum X (resultant) of the two right primary vectors II (coordinates 2, 1) and III (coordinates -1, 1) is not coincident with the left fundamental vector; that is, it does not have coordinates (1, 5). Therefore, we seek values for x and y that will expand or contract each initial primary (vectors II and III), so that the terminal primary vectors add to a resultant that is coincident with the vector representing the matched fundamen-

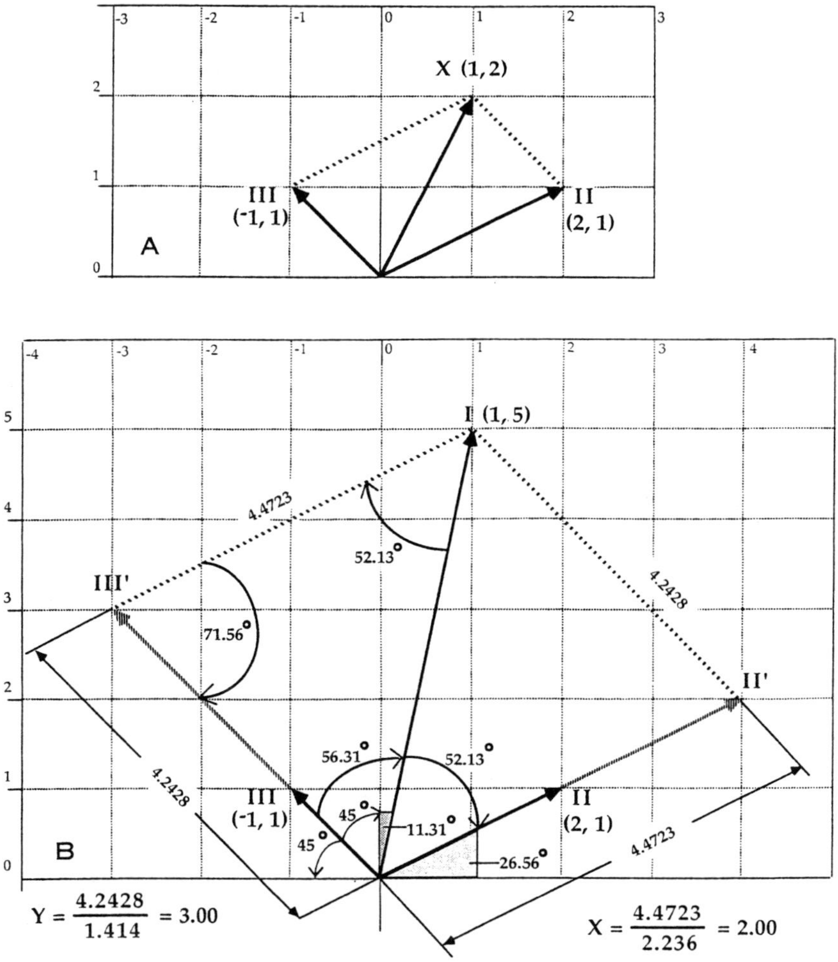

Solution to Simultaneous Equations

Figure 3. Simple simultaneous equations may be solved in vector format.

tal (vector I). The expansion-contraction coefficients are multiplication operators, external to the column vectors; the expansion-contraction is effected by multiplication operators without dimensions and is without units. By themselves, the multiplication operators do not appear as vectors in the Euclidean fundamental color space.

We may calculate x and y from the information in Figure 3B using the usual Pythagorean theorem and the usual trigonometric functions. Immediately, we may compute the lengths of the three vectors. From these, we may calculate all angles using the cosine function. Let I equal the terminal vector with coordinates 1, 5 (length 5.099). Let II equal the length of the initial vector (2.236), and let II′ equal the length of the terminal vector, that is, *x*II. Similarly for III and *y*III′. From trigonometric theorems, 5.099/(sin 71.56°) = II′/(sin 56.31°). Solving, II′ = 4.4723, the length of the terminal vector, remembering that II = 2.236, the length of the initial vector. Thus, $x(2.236) = 4.4723$, and $x = 2$. Similar calculations demonstrate that $y = 3$. Coefficients x and y are not vectors, but multipliers of vectors. That is, the initial length of vector II must be multiplied by 2.00, and vector III must be multiplied by 3.00. We may even have the same initial equation with different multipliers; thus, x may equal 4 instead of 2:

$$\begin{bmatrix}1\\5\end{bmatrix} = 8\begin{bmatrix}0.50\\0.25\end{bmatrix} + 3\begin{bmatrix}1\\1\end{bmatrix}. \tag{6.7}$$

Nevertheless, the plot of the terminal vectors (after multiplication by the tristimulus values) is precisely equivalent to those illustrated in Figure 3B. This example is precisely analogous to the color-matching equation, where the multipliers are the tristimulus values. Note that I solved for the multipliers in this ordinary set of simultaneous equations. In the color-matching experiment, however, the multipliers are obtained experimentally.

The multipliers x and y are simply waiting to perform their multiplication. Given a single terminal vector, we may move any multiplier or tristimulus value outside the column vector; the initial length of the vector is changed to conform to the selected tristimulus value, but the terminal length is always the same. Since the tristimulus values or multipliers are discretionary, the addition of x to y is meaningless. This addition, however, is performed commonly in the calculation of the CIE chromaticity coordinates, as I have indicated.

Figure 4 continues to illustrate the differences between tristimulus values and fundamentals in the color-matching equation. In a color-matching equation, the tristimulus values of matrix **A** are easily confused with the fundamentals of the primaries in matrix **E**. The row entries in matrix **A** weight the fundamentals of the primaries, the column vectors in matrix **E**. The top row illustrates the terminal color-matching equation, where the coefficients are unity and the column vectors are properly modified to achieve the match specified by the column vector on the right (the standard). The fundamentals $\mathbf{N}^*_B$ and $\mathbf{N}^*_G$ and $\mathbf{N}^*_R$ add to $\mathbf{N}^*_C$; this color-matching equation balances.

Figure 4B and C depict exactly the same equation as Figure 4A, although they have very different tristimulus values. In Figure 4B, every element of $\mathbf{N}^*_B$ and $\mathbf{N}^*_G$ and $\mathbf{N}^*_R$ has been divided by 0.333 and arbitrary tristimulus values placed outside the column vectors. When the column vectors are multiplied by tristimulus value 0.333 the column vectors are identical to those in Figure 4A. The

$$
\text{A} \quad 1.\begin{bmatrix} 9. \\ 2. \\ 4. \\ 6. \\ 5. \end{bmatrix}_{N_B^*} + 1.\begin{bmatrix} 4. \\ 6. \\ 2. \\ 3. \\ 5. \end{bmatrix}_{N_G^*} + 1.\begin{bmatrix} 9. \\ 7. \\ 6. \\ 4. \\ 2. \end{bmatrix}_{N_R^*} = \begin{bmatrix} 22. \\ 15. \\ 12. \\ 13. \\ 12. \end{bmatrix}_{N_C^*}
$$

$$
\text{B} \quad 0.333\begin{bmatrix} 27.003 \\ 6.001 \\ 12.001 \\ 18.002 \\ 15.002 \end{bmatrix} +0.333\begin{bmatrix} 12.001 \\ 18.002 \\ 6.001 \\ 9.002 \\ 15.002 \end{bmatrix} +0.333\begin{bmatrix} 27.003 \\ 21.002 \\ 18.002 \\ 12.001 \\ 6.001 \end{bmatrix} = \begin{bmatrix} 22. \\ 15. \\ 12. \\ 13. \\ 12. \end{bmatrix}
$$

$$
\text{C} \quad 0.400\begin{bmatrix} 22.500 \\ 5.000 \\ 10.000 \\ 15.000 \\ 12.500 \end{bmatrix} +0.500\begin{bmatrix} 8.000 \\ 12.000 \\ 4.000 \\ 6.000 \\ 10.000 \end{bmatrix} +0.600\begin{bmatrix} 15.000 \\ 11.667 \\ 10.000 \\ 6.667 \\ 3.333 \end{bmatrix} = \begin{bmatrix} 22. \\ 15. \\ 12. \\ 13. \\ 12. \end{bmatrix}
$$

dimensionless weights

Figure 4. Difference between tristimulus values and fundamentals.

three arbitrary tristimulus values need not be equal to each other; in Figure 4C, 0.400 has been arbitrarily selected to multiply $\mathbf{N}_B^*$, 0.500 has been arbitrarily selected to multiply $\mathbf{N}_G^*$, and 0.600 has been aribitrarily selected to multiply $\mathbf{N}_R^*$. Tristimulus values are merely dimensionless weights. Note that the units belong to the fundamentals, not to the tristimulus values.

Equations 6.6 and 6.7 seem virtually identical to the valid color-matching equation, 6.4. Nevertheless, these equations have a profound difference. In equation 6.6, we are given the column vectors, that is, the fundamentals of the primaries, and we solved for the multipliers, or the tristimulus values. In the color-matching experiment, initially we are ignorant of the primary fundamentals and obtain the tristimulus values experimentally.

Given the pervasiveness of solving linear simultaneous equations for the tristimulus values, colorimetrists have assumed that the multipliers (percentages), emerging from a color-matching experiment, are fundamentals. This is a serious error. The multipliers do not lead to meaningful scalar products, for example, computed only between vectors, and not between their multipliers. The problem demands the derivation of the fundamentals. Simply stated, in comparison to the usual simultaneous equations described in mathematical textbooks, the color problem is inverted.

The question now is, Given the tristimulus values alone, can I solve for the column matrices, the fundamentals? We *can* effect this solution, and that is one aspect of matrix-**R** operations. A fascinating peculiarity of the color processing mechanism permits a precise resolution. We have matrix **A** alone, and we make no assumptions beyond those inherent in Grassmann's laws.

The Color-Matching Experiment

Maximum Saturation Method

In its most elegant format, an observer matches the sensation evoked by one passband of the visible spectrum (one monochromatic stimulus at unit irradiance, termed the *variable light*) on the left side of a bipartite field with the compounded sensation evoked by the additive mixture of three independent passbands of the visible spectrum (three monochromatic stimuli termed primaries) on the right side of a bipartite field. Except for the instance where the stimulus on the left is identical to one of the primaries on the right (discussed in a later section), the observer will report that the match is not possible. One predetermined primary (not *any* primary) is then transferred from right to left, and the observer adjusts the magnitude of each primary—the primary on the left now mixed with the unit variable stimulus on the left, and the two primaries mixed with each other on the right. Eventually the observer announces that a match has been achieved, and the proportions of the primaries, termed *tristimulus values*, are recorded as one row of matrix **A**. The primary that has been shifted from right to left assumes a negative value. Eventually the *k* rows of matrix **A** are complete, and the experiment is concluded.

The primary to be transferred from the right to the left is dictated solely by the variable stimulus currently under investigation. Taking the spectrum in wavelength order, when the variable wavelength is shorter than the blue primary B, the variable stimulus in the left field is combined with the green primary G. The blue primary B and the red primary R are combined in the right field.

As the variable wavelength passes through the blue primary, the test field (the left field), now containing one unit of variable light equivalent (in intensity and spectral composition) to one unit of blue primary, may be matched by a combination of the three primaries, in fact by 1, 0, 0 units of these primaries in BGR order. No transfer of a primary is necessary at those three wavelengths where the variable wavelength is identical to one of the primary wavelengths.

As the variable wavelength passes to wavelengths longer than the blue primary B, the red primary R is transferred to the left field. Hence the primary transferred acquires a negative value. This is supported by Grassmann's additivity law, which says that given two matching lights, the addition of matching lights simultaneously to both lights preserves the match. The lights appear to be different from before the addition, but the match is preserved. The match is actually made as

$$1C + r_t R = g_t G + b_t B, \tag{6.8}$$

where the notation is that of equation 6.2, but $-r_t R$ is added to both sides of this color-matching equation and the variable light is matched by the three primaries according to that equation:

$$1C = -r_t R + g_t G + b_t B. \tag{6.9}$$

The tristimulus values of the spectral lights at each variable wavelength are now loaded wavelength for wavelength, row by row, into matrix **A**.

Maxwell-Matching Procedure

In the so-called Maxwell-matching color-matching experiment (Maxwell, 1872), a reference white is set up in one field, say, the right field. The three primaries are combined in a triplet on the left side of the bipartite field. The three primaries are adjusted in magnitude by the observer until he declares a match has been achieved. The color-matching equation that describes this match is:

$$1W_C = r_t R + g_t G + b_t B. \tag{6.10}$$

This equation says that one standard unit of reference white is matched by r_t, the decimal fraction of the standard amount of red primary R, additively mixed with g_t, the decimal fraction percentage of the standard amount of the green primary G with b_t, the decimal fraction of the standard amount of the blue primary B. The standard amount may be arbitrarily chosen by the experimenter, but a convention often used is that one unit of R combined with one unit of G and one unit of B match the reference white. The tristimulus values of equation 6.9 would be all ones on the right-hand side under such conditions of standardization.

Thereafter, a variable waveband is chosen for investigation as in the maximum saturation method. The variable waveband replaces the primary nearest to it in the spectrum, so that the reference white is always being matched with red, green, and blue components, of which two are primaries and one a variable wavelength at any time. The variable wavelength is thus gauged by

$$1W_C = r_t' R' + g_t G + b_t B, \tag{6.11}$$

where the prime symbol signifies that the red primary has been replaced with a red variable light.

Since the reference white is the same in each trial as in every other trial, Grassmann's law of transitivity may be invoked to set the right-hand sides of equations 6.10 and 6.11 equal to each other. Grassmann's transitivity law says that two lights, each of which matches a third light, match each other. So

$$1R + 1G + 1B = 1W_C = r'_t R' + g_t G + b_t B \tag{6.12}$$

$$r'_t R' = 1R + (1 - g_t)G + (1 - b_t)B \tag{6.13}$$

$$1R' = [1/r'_t]R + [(1 - g_t)/r'_t]G + [(1 - b_t)/r'_t]B. \tag{6.14}$$

This equation says that the tristimulus values of a unit quantity of the variable red stimulus are the expressions in square brackets on the right-hand side of equation 6.14. The authority for dividing both sides of equation 6.13 by r'_t is Grassmann's proportionality law, which is a corollary of the additivity law. The proportionality law states that two lights that match may be multiplied by the same constant and will still match.

Equations 6.12 through 6.14 show why it has been the colorist's driving ambition over the course of time to set the units of the primaries so that unit quantities match the reference white. Although this is not necessary to the analysis, it vastly simplifies the coefficients of the equations employed.

7

Matrix **A**, Matrix **E**, and the Definition of Matrix **R**

Matrix R as the Instrument of the Wyszecki Hypothesis

Matrix **A** has a conspicuous characteristic, because three rows may, in general, correspond to the three primaries, and indeed, the stimuli represented by these rows are identical to the stimuli represented by the primaries. Therefore, these three rows comprise a 3×3 identity matrix. The first row, corresponding to the first primary, is matched by the coefficient one multiplied by the first primary, the coefficient zero multiplied by the second primary, and the coefficient zero multiplied by the third primary. The same is true for matching the second and third rows corresponding to the primaries. In fact, an identity matrix identifies the rows in an **A** matrix that correspond to the primaries. If no rows correspond to the primaries, as may be the case when the primaries are compounded or imaginary (for examples with CIE X,Y,Z), then the 3×3 identity matrix representing the primaries may be augmented to the **A** matrix.

The primaries of matrix **A** may be transformed linearly to other primaries in other **A** matrices; the same is true for **E** matrices. These and other transformations will be discussed.

Matrix **Q**, 3×1, is the well-known tristimulus values referring to any radiometric function with respect to a triplet of specified color-matching functions, say, matrix **A**. By definition,

$$\mathbf{A'N} = \mathbf{Q}. \tag{7.1}$$

Matrix $\mathbf{Q}$ is a 3×1 column matrix, while the tristimulus values of a monochromatic stimulus in matrix $\mathbf{A}$ are three entries in one row. This may be confusing; nevertheless, the definition for tristimulus values remains valid. The rows of matrix $\mathbf{A}$ represent monochromatic standards—radiometric functions that are column matrices designated $\mathbf{N}$, with an entry of unity in one cell and zeros elsewhere. Then, $\mathbf{A'N} = \mathbf{Q}$, a 3×1 matrix of tristimulus values.

Matrix $\mathbf{N}^*$, $k \times 1$, and as defined above, is the scalar fundamental of any radiometric function $\mathbf{N}$. Recall that the fundamental alone transmits all colorimetric information to the sensorium, to evoke the color sensation. When $\mathbf{N}^*$ is the fundamental of $\mathbf{N}$, then

$$\mathbf{A'N} = \mathbf{A'N}^* = \mathbf{Q}. \tag{7.2}$$

This equation states that the tristimulus values of a radiometric function are identical to the tristimulus values of its fundamental.

Invariably,

$$\mathbf{A'B} = \mathbf{0}. \tag{7.3}$$

This equation states that the tristimulus values of any residual are 0, 0, 0, evoking a black sensation (black) with respect to any triplet of primary color stimuli. This is more than an assertion and will be proved later.

Now,

$$\mathbf{A'E} = \mathbf{I} \tag{7.4}$$

and

$$\mathbf{E'A} = \mathbf{I}, \tag{7.5}$$

where $\mathbf{I}$ is a 3×3 identity matrix. This states that the tristimulus values of three primaries in a color-matching experiment are (1, 0, 0), (0, 1, 0), and (0, 0, 1) computed with the color-matching functions emerging from that experiment. We have already examined the logic underlying these critical relationships.

Define:

$$(\mathbf{A'A})^{-1} = (\mathbf{M}_a)^{-1} \tag{7.6}$$

and define

$$(\mathbf{E'E})^{-1} = (\mathbf{M}_e)^{-1}. \tag{7.7}$$

Now,

$$(\mathbf{A'A})(\mathbf{A'A})^{-1} = \mathbf{I}, \tag{7.8}$$

but

$$\mathbf{A'E} = \mathbf{I}. \tag{7.9}$$

Then

$$\mathbf{A'E} = (\mathbf{A'A})(\mathbf{A'A})^{-1}. \tag{7.10}$$

Therefore,

$$\mathbf{A} = \mathbf{E}(\mathbf{E'E})^{-1} = \mathbf{E}(\mathbf{M}_\mathrm{e}^{-1}), \tag{7.11}$$

and similarly,

$$\mathbf{E} = \mathbf{A}(\mathbf{A'A})^{-1} = \mathbf{A}(\mathbf{M}_\mathrm{a}^{-1}). \tag{7.12}$$

So, given matrix **A**, a triplet of color-matching functions, we may immediately compute the corresponding matrix **E**—the triplet of fundamentals corresponding to the primaries. Analogously, given a matrix **E**, a triplet of fundamentals corresponding to the primaries, we may immediately compute matrix **A**, the color-matching functions that would emerge in a color-matching experiment using the triplet of primaries in matrix **E**. Note that we cannot compute $\mathbf{N}_p$, the radiometric functions of the primaries, but we may compute an infinite number of metamers to these primaries by the addition of residuals.

With respect to any matrix **A**, the sums of the three columns are the respective tristimulus values for an equal-energy source. Stated differently, the sum of each column is the dimensionless number that weights the fundamental of the respective primary. This is true because each row of matrix **A** is the tristimulus value of one monochromatic stimulus spanning the visible spectrum, and the sums of the tristimulus values are the tristimulus values of the sum of all unit-power monochromatic stimuli—an equal-energy stimulus. The CIE desired equal tristimulus values to specify the sensation evoked by an equal-energy source. This was accomplished, presumably, by making the sums of the columns almost equal; nothing prevented the CIE from making the sums exactly equal because the adjustment of the sums is irrelevant. For example, the X column might have been doubled, and the sum would also have been doubled. Nevertheless, the fundamentals in the corresponding **E** matrix would be half the magnitude, but in the valid color-matching equation, they would have been weighted by tristimulus value for X multiplied by 2. The final fundamental for primary X therefore would be precisely the same; the **M** matrices, defined above, are automatically compensatory.

Clearly,

$$\mathbf{M}_\mathrm{a} = \mathbf{M}_\mathrm{e}^{-1}, \tag{7.13}$$

and

$$\mathbf{M}_\mathrm{e} = \mathbf{M}_\mathrm{a}^{-1}. \tag{7.14}$$

Define $k \times k$ matrix **R**:

$$\mathbf{R} = \mathbf{A}(\mathbf{M}_\mathrm{a})^{-1}\mathbf{A'} = \mathbf{A}(\mathbf{A'A})^{-1}\mathbf{A'}. \tag{7.15}$$

But $\mathbf{E} = \mathbf{A}(\mathbf{M}_a)^{-1}$, and therefore,

$$\mathbf{R} = \mathbf{E}\mathbf{A}'. \tag{7.16}$$

By equation 7.16, the fundamental of any monochromatic color stimulus is obtained by weighting the fundamentals of the primaries by the tristimulus values in the corresponding row of matrix **A**. The columns (or rows) of matrix **R** are therefore the fundamentals of the spectral stimuli; several examples will be demonstrated. Similar operations give

$$\mathbf{R} = \mathbf{A}\mathbf{E}'. \tag{7.17}$$

Therefore, we have four powerful equations relating matrices **A**, **E**, and **R**:

$$\mathbf{A}'\mathbf{E} = \mathbf{I},\ \mathbf{E}'\mathbf{A} = \mathbf{I},\ \mathbf{E}\mathbf{A}' = \mathbf{R},\ \mathbf{A}\mathbf{E}' = \mathbf{R}. \tag{7.18}$$

Matrix **R** is the basic construct of modern colorimetry. Matrix **R** is invariant, as will be proved, and the identical matrix **R** may be derived from any **A** matrix or its linear transform or any **E** matrix or its linear transform. Figure 5 portrays the characteristics of matrix **R** with several computer renderings. Figure 5A displays an overview together with a two-dimensional projection. Each column (or row) is the scalar fundamental of a monochromatic stimulus, ranging from 380 nm to 770 nm. Each fundamental is the locus of scalar products between the given monochromatic stimulus and all other monochromatic stimuli. The entries peak at three summital modes, the greatest magnitude in the blue portion of the spectrum. The orthographic projection in A also makes this demonstration. Figure 5B, is a side view, displaying the negative values of matrix **R**. Each spectral fundamental (row or column) spans the visible spectrum, and all spectral fundamentals have some negative values. Matrix **R** defines the vector configuration of the fundamental color space; each entry is a scalar product that defines the vector lengths of monochromatic stimuli and their intervening angles. The identical matrix **R** may always be computed from any independent triplet of primary fundamentals or from their independent color-matching functions.

Matrix **R** is symmetrical. Each row (or column) is the scalar fundamental of a monochromatic, spectral color stimulus. Figure 6 illustrates several selected monochromatic fundamentals, each a row or column in matrix **R** (see Figure 5) and is the locus of scalar products in the fundamental color space. Figure 6A depicts the fundamentals of 390 nm and 420 nm, Figure 6B depicts the fundamentals of 450 nm and 470 nm, Figure 6C depicts the fundamentals of 480 nm and 520 nm, Figure 6D depicts the fundamentals of 570 and 600 nm, and Figure 6E depicts the fundamentals of 640 nm and 680 nm.

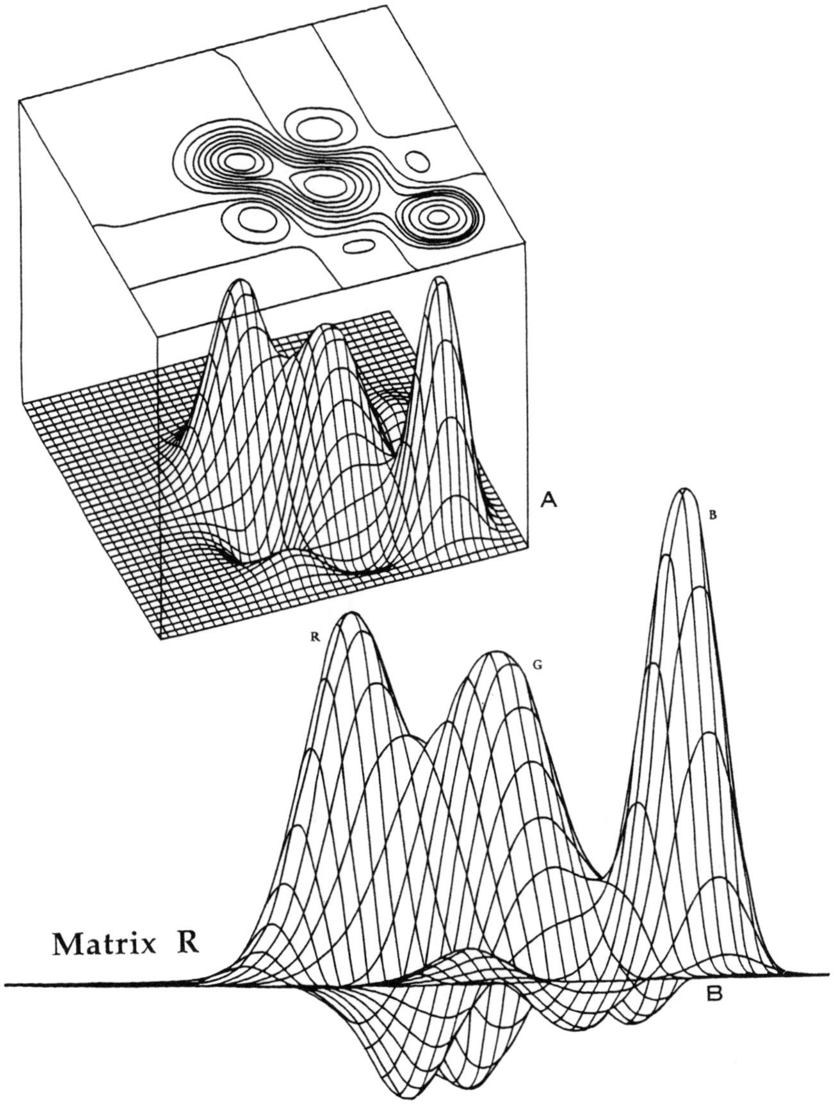

Figure 5. Two computer renderings of matrix **R**, the projection operator.

Every monochromatic fundamental spans the entire visible spectrum, therefore overlapping with the others. All fundamentals tend to peak at one or another of the summital wavelengths, and all have some negative values. Every

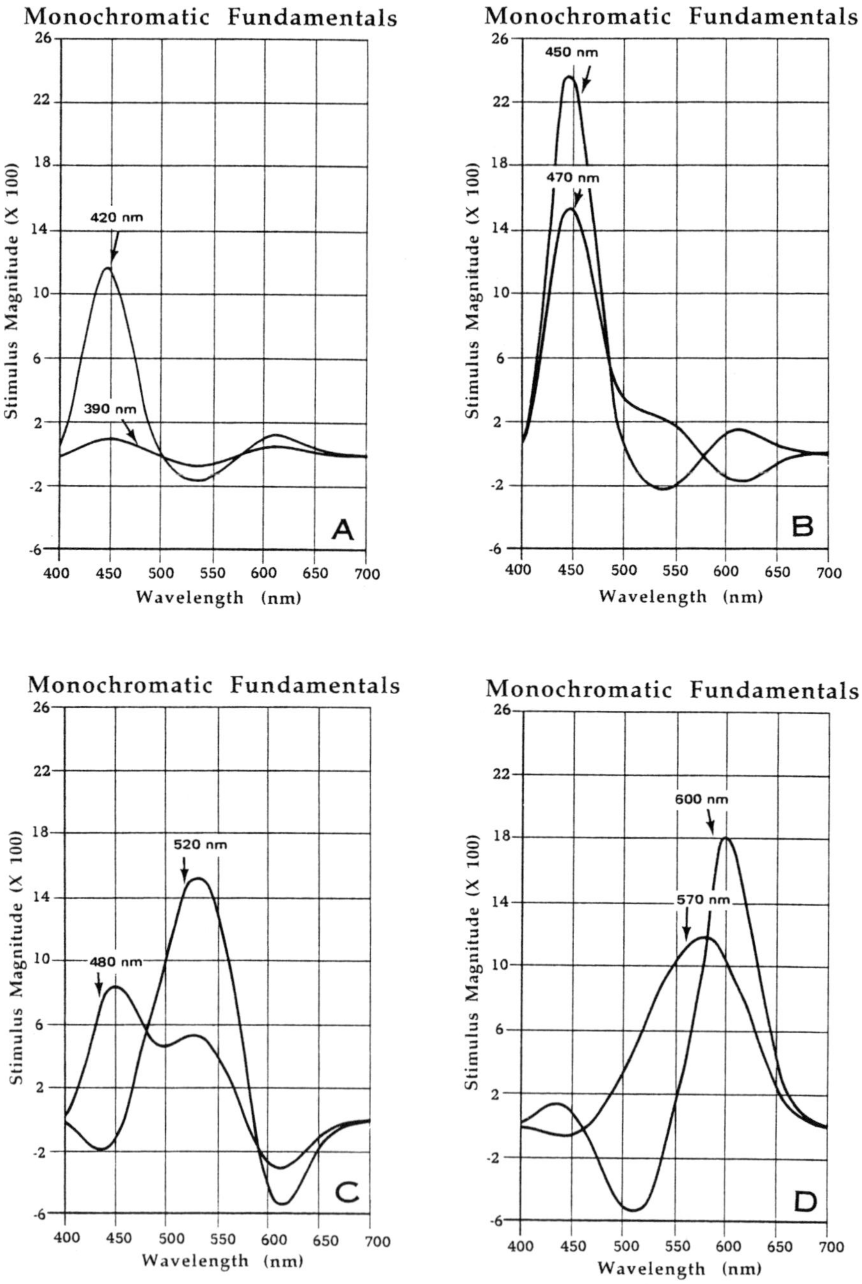

Figure 6. Explicit renderings of selected monochromatic fundamentals. *Continued next page.*

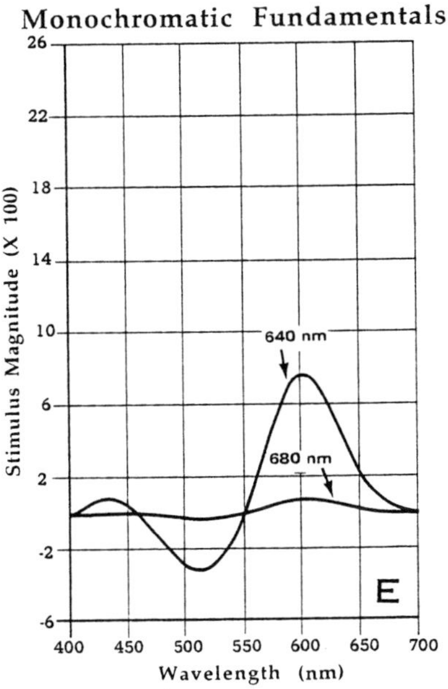

Figure 6. Continued.

fundamental is a linear combination of the weighted columns (or rows) of matrix **R**.

As shown in Figures 5 and 6, matrix **R** exhibits three maxima. The first maximum is approximately 443–447 nm, the second is approximately 531–535 nm, and the third is approximately 595–600 nm. The wavelengths of these maxima, termed *summital wavelengths*, have appeared and reappeared during the long history of color science. They emerged in photometric experiments concerning complementary color stimuli by MacAdam (1938) and have been predicted using theoretical considerations by Kuehni (1979) and Thornton (1972). By the reasoning of Bird and Jones (1965), they foretell the probable characteristics of human cone functions. The summital wavelengths are the peaks of the CIE B,G,R color-matching functions for both the 1931 and 1964 standard observers (see Wyszecki & Stiles, 1982). In the design of color films, the summital wavelengths are known as the Maxwell-Ives maxima, after James Clerk Maxwell and Herbert E. Ives, who noted their consistent recurrence. Eventually they were calculated by Neugebauer (1956), who set out to calculate the least-square fit of the equal-energy spectrum taken a monochromat at a time to the Maxwell-Ives criterion. The summital wavelengths proved to be the maxima of the resulting best fit. The summital wavelengths remain important in commercial applica-

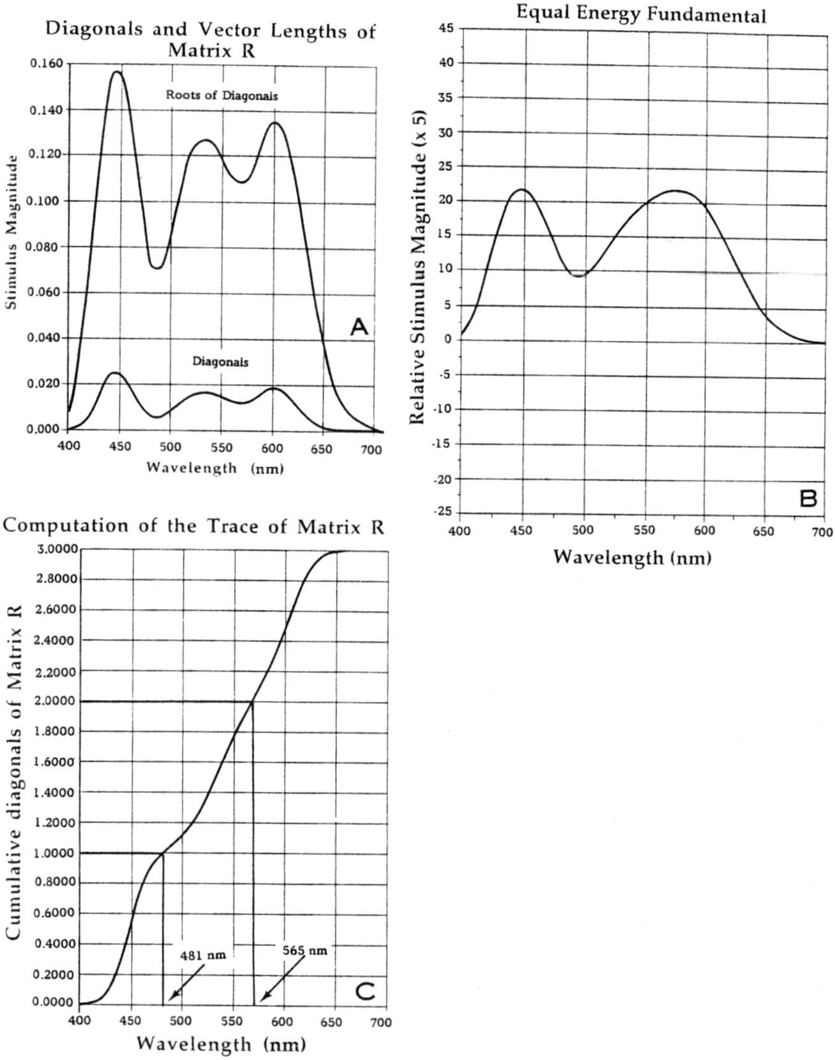

Figure 7. Three properties of matrix **R**, shown in Figure 5.

tions (see Hardy & Wurzburg, 1937; Hanson & Brewer 1954a, 1954b; Ross, 1970).

Figure 7 illustrates three additional properties of matrix **R**. Figure 7A plots the diagonal entries in matrix **R**—the squares of the monochromatic vector lengths on the envelope of vector **R**—and also the roots of the diagonal entries; the roots are the magnitudes of the vectors representing the monochromatic stimuli in fundamental color space. The diagonal entries of matrix **R** peak at precisely the summital wavelengths. Figure 7B plots the sums of columns (or rows) of matrix **R** and is equivalent to matrix **R** multiplying the column matrix representing an equal-energy stimulus where every entry is unity. It demonstrates the characteristic bimodal appearance of fundamentals representing desaturated stimuli; this fundamental, with its two modes of almost identical magnitude, is the fundamental of an equal-energy source. Note here the two peaks of virtually identical magnitude; this function occurs repeatedly in studies of the fundamental color space. Figure 7C is the cumulative sum of the diagonal entries of matrix **R**; the total sum, termed the *trace*, is precisely equal to three, the dimensionality of matrix **R** and, of course, the dimensionality of fundamental color space. This suggests a tripartite division of the visible spectrum where the "trace values" are 1.00 and 2.00. (See also Figures 14 and 15, which depict the invariant vector structure of the fundamental color space.) Later we shall examine further the demonstrations in Figure 7.

Define:

$$\mathbf{R}_{\mathrm{B}} = \mathbf{I} - \mathbf{R}, \qquad (7.19)$$

where matrix **I** is a $k \times k$ identity matrix and matrix **R** is defined above. The entries in matrix $k \times k$ $\mathbf{R}_{\mathrm{B}}$ are the residuals of every monochromatic stimulus, each corresponding to one passband in the visible spectrum. Matrix $\mathbf{R}_{\mathrm{B}}$ is symmetrical. Recall that $\mathbf{B} = \mathbf{N} - \mathbf{N}^*$. With respect to equation 7.19, each matrix **N** corresponds to a monochromatic stimulus with unity in the row (or column) representing the same monochromatic stimulus and zeros elsewhere, so that the k**N** matrices collectively form a $k \times k$ identity matrix.

Figure 8 illustrates several rows (or columns) of matrix $\mathbf{R}_{\mathrm{B}}$, the residuals of monochromatic stimuli. Every residual spans the visible spectrum, and all have negative values. All monochromatic residuals are computed by subtracting matrix **R** from an identity matrix of the same order. Each diagonal element in the identity matrix represents the unit irradiance of the monochromatic radiometric funtion. Thus, the monochromatic residuals are equal to the monochromatic fundamentals, except for a sign change and the diagonal elements. In the matrix of monochromatic residuals, the diagonal element equals unity minus the square of the vector length, the diagonal entry in matrix **R**. This accounts for the enormous spikes at the wavelength of each monochromatic stimulus, corresponding to the diagonals in matrix $\mathbf{R}_{\mathrm{B}}$ and in matrix **R**. The residual of every color stimulus is a linear combination of the weighted columns (or rows) in matrix $\mathbf{R}_{\mathrm{B}}$. Monochromatic residuals have exactly the same properties as all other residuals; even with their spikes, their tristimulus values are invariably 0, 0, 0.

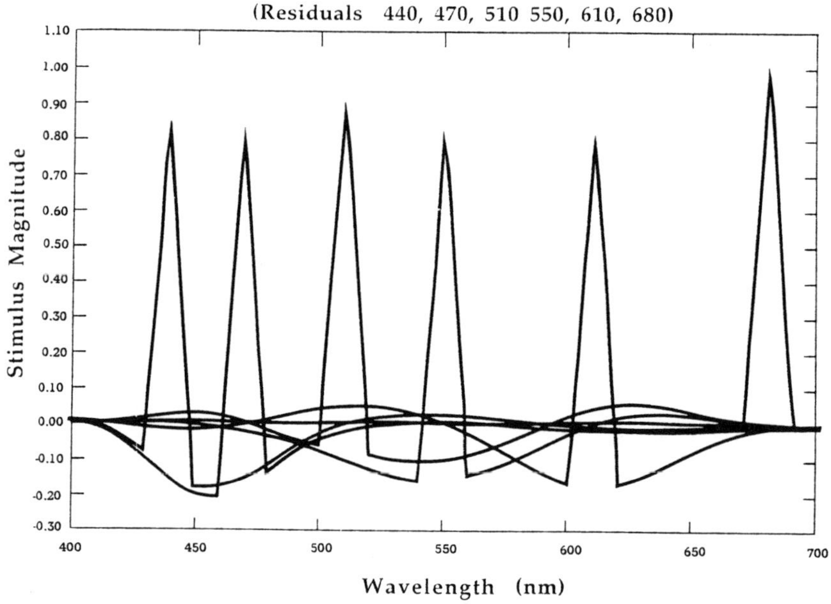

Figure 8. Residuals of several selected monochromatic stimuli.

Matrix R as a Resolution Matrix

Matrix **R** has the extraordinary capability to resolve directly any radiometric function into its fundamental and residual, as demanded by the Wyszecki hypothesis. Several of these resolutions were demonstrated in Figure 1.

Because matrix **R** comprises the fundamentals of ordered spectral color stimuli at unit irradiance, every fundamental is the sum of appropriately weighted columns of matrix **R**. Thus, given any radiometric function **N**, the fundamental **N*** is easily computed:

$$\mathbf{RN} = \mathbf{N}^*. \tag{7.20}$$

I can now show that the tristimulus values of radiometic **N** are identical to those of its fundamental **N***.

By substituting equation 7.15 into equation 7.20,

$$\mathbf{A}(\mathbf{A}'\mathbf{A})^{-1}\mathbf{A}'\mathbf{N} = \mathbf{N}^*. \tag{7.21}$$

Multiply by A′:

$$\mathbf{A}'\mathbf{A}(\mathbf{A}'\mathbf{A})^{-1}\mathbf{A}'\mathbf{N} = \mathbf{A}'\mathbf{N}^*. \tag{7.22}$$

Thus,

$$\mathbf{I}\mathbf{A}'\mathbf{N} = \mathbf{A}'\mathbf{N} = \mathbf{A}'\mathbf{N}^*. \tag{7.23}$$

Since $\mathbf{A}'\mathbf{N} = \mathbf{A}'\mathbf{N}^*$, the tristimulus values for **N** and **N*** are identical.

In a sense, the elements in a radiometric function become k multistimulus values, each modulating the fundamental of one monochromatic stimulus at unit irradiance in the visible spectrum, analogous to three tristimulus values, each modulating the fundamental of one primary. When so considered, multistimulus values and tristimulus values are merely multipliers operating on fundamentals. The units belong to the fundamentals.

As a corollary,

$$\mathbf{RN^*} = \mathbf{N^*}. \tag{7.24}$$

From equation 7.15,

$$\mathbf{A(A'A)^{-1}A'N^*} = \mathbf{N^*}. \tag{7.25}$$

Premultiply by $\mathbf{A}'$:

$$\mathbf{A'A(A'A)^{-1}A'N^*} = \mathbf{A'N^*}. \tag{7.26}$$

Thus,

$$\mathbf{IA'N^*} = \mathbf{A'N^*}. \tag{7.27}$$

Figures 9A and 10A plot radiometric functions of Munsell Y 8/12 and R 5/12 viewed under an equal-energy illuminant as measured by Dorothy Nickerson (personal communication, 1955). As with all other real color stimuli, all values are positive. These functions have been premultiplied by matrix **R** to give the fundamentals shown in Figures 9B and 10B. Fundamental Y 8/12 has all positive values, characteristic of color stimuli with low saturation (compare with the fundamentals of spectral stimuli shown in Figure 6). Fundamental R 5/12 has some negative values, characteristic of color stimuli with higher saturation (compare with the fundamentals in Figure 9 and the fundamentals of spectral stimuli shown in Figure 6). The bimodal distribution of the equal-energy fundamental is the sum of the columns (or rows) of matrix **R** and was illustrated in Figure 7B. Figures 9C and 10C are the residuals obtained by subtracting the fundamental from the radiometric function. These residuals display four crossovers, typical of radiometric functions of color stimuli found in natural surroundings. The residuals have both positive and negative values, a characteristic of all residuals.

The bimodal distributions associated with the fundamentals of desaturated stimuli may appear strange to modern colorimetrists, but they are quite old. They were inherent in the work of Herman von Helmholtz; Peddie (1922) and Hecht (1924) have operated on data recorded by Helmholtz (1852) to obtain the functions shown in Figure 11.

We may now compute the fundamentals of the radiometric functions used as primaries in the color-matching experiment. Gather the radiometric functions

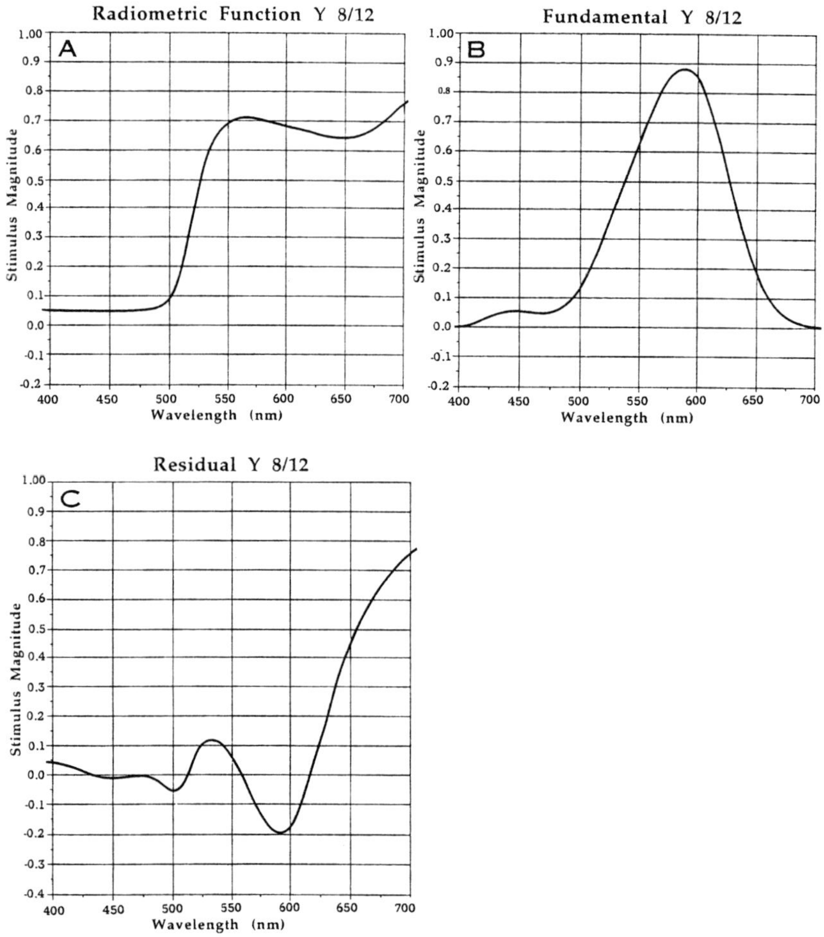

Figure 9. Demonstration of matrix **R** as a resolution matrix, mimicking the color processing mechanism and fulfilling the Wyszecki hypothesis.

into a $k \times 3$ matrix $\mathbf{N}_\mathrm{p}$. Then, by equation 7.24,

$$\mathbf{RN}_\mathrm{p} = \mathbf{E}. \tag{7.28}$$

Matrix $\mathbf{N}_\mathrm{p}$ may then be substituted in the color-matching equation, 6.4, or an **A** matrix may be computed from matrix **E**.

This resolution property of matrix **R** will now receive additional consideration.

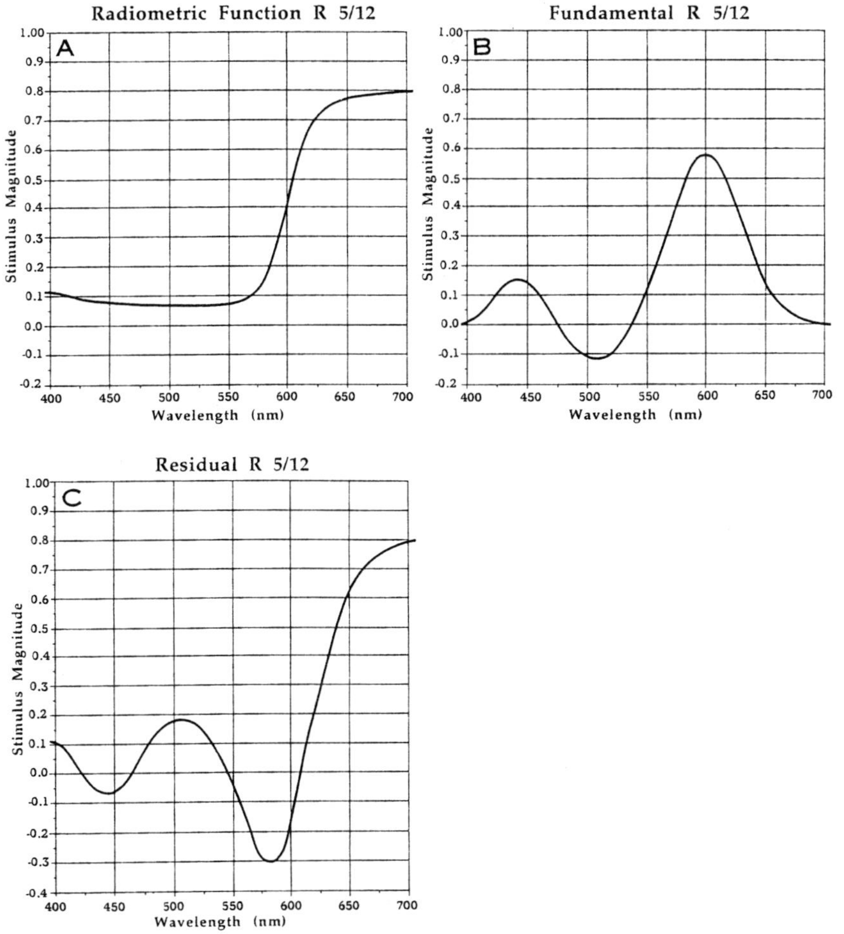

Figure 10. Second demonstration of matrix **R** as a resolution matrix, mimicking the color processing mechanism and fulfilling the Wyszecki hypothesis.

Matrix R as an Orthogonal Projector

Inherent Limitations of the Visual Color Processing Mechanism

Momentarily, give an anthropomorphic interpretation to the color processing mechanism (similar to Maxwell's demon; see Cohen, 1988), burdened with serious maledictions. The color processing mechanism has only three "channels" and can transmit only three dimensions of an incoming k-dimensional color stimulus. The long-suffering color processing mechanism, knowing that full

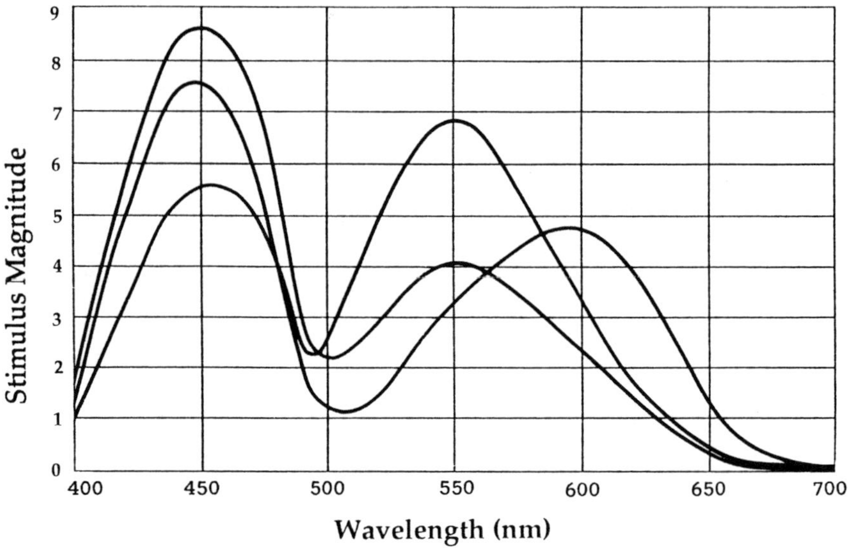

Figure 11. Bimodal fundamental functions, characteristic of color stimuli with low saturations, isolated with matrix **R** operations.

transmission is far beyond its capability, suffers massive anxiety attacks. Given its deficiencies, the mechanism transmits only a selected portion of the stimulus, representing the maximum possible information, and simultaneously ignores or rejects the remaining information. As reported in Cohen (1988), the color processing mechanism does its level best.

The best, however is not good enough, and penalties are exacted. The penalty is the occurrence of metamers, where the identical fundamental is extracted from two very different radiometric functions. The color sensation evoked by one metamer is precisely identical to the color sensation evoked by another, although the radiometric functions are very different; one metamer may be substituted for the other in the additive compounding of any color stimulus. Metamers occur because some information, the residual, is ignored and lost. Metamers represent failures of the color processing mechanism, where it gives identical interpretations to vastly dissimilar stimuli, and so fails to differentiate among stimuli belonging to the same metameric suite.

These deficiencies have been exploited. Indeed, if the visual system completely processed every radiometric function to evoke a different color sensation, then contemporary color printing, color photography, and color television would be virtually nonexistent. Modern color reproduction is essentially the generation of new color stimuli metameric to the original color stimuli.

The Color Processing System and Matrix **R**, the Projection Operator

Here we explore mathematics that describe and even mimic the error-minimizing capability of the visual system.

The equations

$$\mathbf{A}(\mathbf{A}'\mathbf{A})^{-1}\mathbf{A}'\mathbf{N} = \mathbf{N}^* \tag{7.29}$$

and

$$\mathbf{A}'\mathbf{B} = 0 \tag{7.30}$$

are closely associated. We may demonstrate that the first equation minimizes mathematical prediction error and prove the second equation, or we may demonstrate that the second equation is a condition of the Wyszecki hypothesis and prove the first equation; Cohen and Kappauf (1985) employed the first proof considered here.

In practice, the radiometric space is k-dimensions, the fundamental color space is three dimensions, and the residual space is $(k - 3)$-dimensions. Figure 12 illustrates the essential geometry in artificial spaces that may be visualized. In Figure 12, a three-space represents the radiometric functions of color stimuli, a plane represents the fundamentals of color stimuli in a two-dimensional subspace, and an orthogonal line represents the residuals in a one-dimensional subspace. The dimensionality of the radiometric space equals the sum of the subspace dimensionalities. The subspaces are always orthogonal to each other; in mathematical terminology, the subspaces are *complementary* (not to be confused with complementary color stimuli). Every vector in each subspace is orthogonal to every vector in the other subspace.

In Figure 12, four vectors—$\mathbf{N}_a$, $\mathbf{N}_b$, $\mathbf{N}_c$, and $\mathbf{N}_d$—represent four radiometric functions in three-dimensional radiometric space. The color-processing mechanism is restricted to transmission of color stimuli in fundamental color space. The radiometric functions are not in the fundamental color space, and therefore, the color processing mechanism effectively projects these **N** vectors orthogonally onto their corresponding $\mathbf{N}^*$ vectors in the fundamental space, minimizing the error between the **N** vector in radiometric space and the corresponding $\mathbf{N}^*$ vector in fundamental color space.

The orthogonal projectors, **R** and $\mathbf{R}_B$, resolve each radiometric vector into two components, the one orthogonal to the other. Vector $\mathbf{N}_a$, for example, is projected orthogonally onto vector $\mathbf{N}_a^*$ in the fundamental color space, because the error in the distance between $\mathbf{N}_a$ and $\mathbf{N}_a^*$ is minimized. If other than orthogonal, the distance is not minimized. Similarly, $\mathbf{R}_B$ orthogonally projects vector $\mathbf{N}_a$ onto the vector $\mathbf{B}_a$ in residual color space.

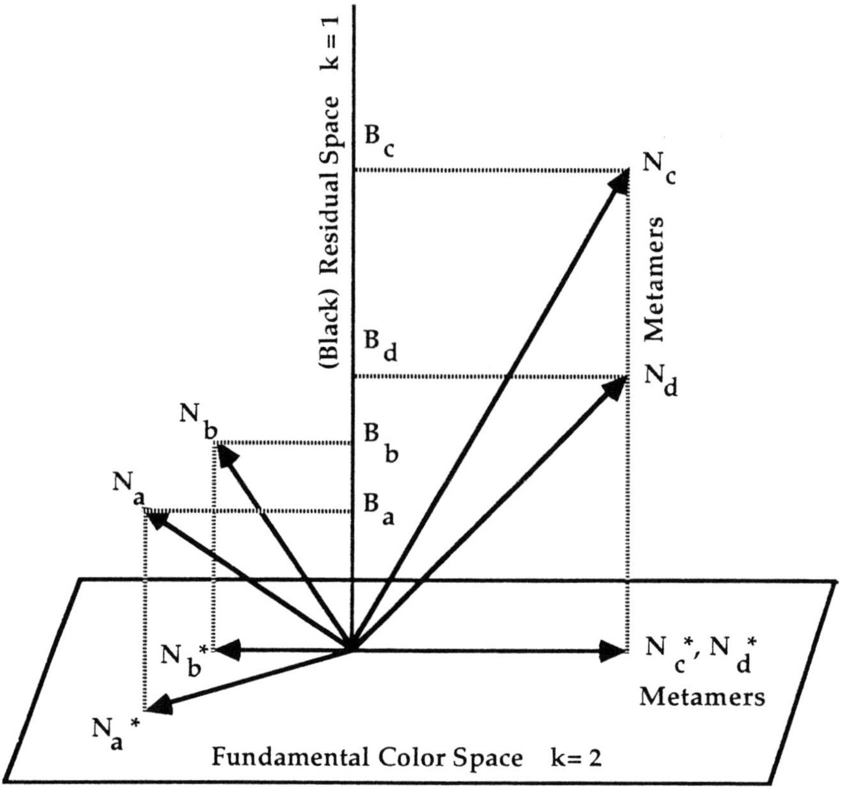

Figure 12. Vector projections onto fundamental color space.

Vectors $\mathbf{N}_c$ and $\mathbf{N}_d$ represent metameric stimuli. These vectors project onto the same vector in fundamental color space, but onto different vectors in residual space. We derive orthogonal projectors, **R** and $\mathbf{R}_B$. We shall prove that the orthogonal projectors are

$$\mathbf{R} = \mathbf{E}(\mathbf{E}'\mathbf{E})^{-1}\mathbf{E}' = \mathbf{A}(\mathbf{A}'\mathbf{A})^{-1}\mathbf{A}' \tag{7.31}$$

and

$$\mathbf{R}_B = \mathbf{I} - \mathbf{R} = \mathbf{I} - \mathbf{E}(\mathbf{E}'\mathbf{E})^{-1}\mathbf{E}' = \mathbf{I} - \mathbf{A}(\mathbf{A}'\mathbf{A})^{-1}\mathbf{A}'. \tag{7.32}$$

The entity $\mathbf{E}(\mathbf{E}'\mathbf{E})^{-1}\mathbf{E}'\mathbf{N}$ minimizes the distance between **N** (in the radiometric space) and **N*** (in the fundamental color space). The fundamental **N*** is a linear combination of the vector fundamentals in the fundamental color space—say, those represented in matrix **E**.

Therefore, we seek a 3 × 1 weighting or transformation matrix **U**, such that

$$\mathbf{EU} = \mathbf{N}^*, \tag{7.33}$$

or, written explicitly in expanded form,

$$\begin{aligned} e_{11}u_1 + e_{12}u_2 + e_{13}u_3 &= n_1^* \\ e_{21}u_1 + e_{22}u_2 + e_{23}u_3 &= n_2^* \\ e_{31}u_1 + e_{32}u_2 + e_{33}u_3 &= n_3^* \\ e_{41}u_1 + e_{42}u_2 + e_{43}u_3 &= n_4^* \\ \cdot \qquad \cdot \qquad \cdot \quad &\quad \cdot \\ e_{k1}u_1 + e_{k2}u_2 + e_{k3}u_3 &= n^*{}_k, \end{aligned} \tag{7.34}$$

where the *e*s are the elements of matrix **E** (three fundamentals in fundamental color space), the *u*s are the elements of matrix **U**, the weighting matrix, and the n^*s are the elements of matrix $\mathbf{N}^*$, the projection of matrix **N** into the fundamental color space.

The distance from **N** (in radiometric space) to $\mathbf{N}^*$ (in fundamental color space) is to be minimized. Stated differently, we minimize $\mathbf{D} = \mathbf{N} - \mathbf{EU}$, equivalent to minimizing the sum of squares, **SS**:

$$\begin{aligned} \mathbf{SS} = \; & (n_1 - [e_{11}u_1 + e_{12}u_2 + e_{13}u_3])^2 \\ & + (n_2 - [e_{21}u_1 + e_{22}u_2 + e_{23}u_3])^2 \\ & + (n_3 - [e_{31}u_1 + e_{32}u_2 + e_{33}u_3])^2 \\ & + (n_4 - [e_{41}u_1 + e_{42}u_2 + e_{43}u_3])^2 \\ & \quad \cdot \qquad \cdot \qquad \cdot \qquad \cdot \\ & + (n_k - [e_{k1}u_1 + e_{k2}u_2 + e_{k3}u_3])^2. \end{aligned} \tag{7.35}$$

Abbreviating,

$$\mathbf{SS} = \mathrm{d}_1{}^2 + \mathrm{d}_2{}^2 + \mathrm{d}_3{}^2 + \mathrm{d}_4{}^2, \ldots, \mathrm{d}_k{}^2. \tag{7.36}$$

I write partial derivatives with respect to u_1, u_2, and u_3, setting the values equal to zero:

$$\begin{aligned} \partial \mathbf{SS}/\partial u_1 &= -2e_{11}d_1 - 2e_{21}d_2 - 2e_{31}d_3, \ldots, -2e_{k1}d_k = 0 \\ \partial \mathbf{SS}/\partial u_2 &= -2e_{12}d_1 - 2e_{22}d_2 - 2e_{32}d_3, \ldots, -2e_{k2}d_k = 0 \\ \partial \mathbf{SS}/\partial u_3 &= -2e_{13}d_1 - 2e_{23}d_2 - 2e_{33}d_3, \ldots, -2e_{k3}d_k = 0. \end{aligned} \tag{7.37}$$

These expressions are a three-element null vector. Disregarding all −2s, the vector is represented as

$$\mathbf{E}'\mathbf{D} = 0. \tag{7.38}$$

However,

$$\mathbf{D} = \mathbf{N} - \mathbf{EU}. \tag{7.39}$$

Therefore,

$$\mathbf{E}'(\mathbf{N} - \mathbf{EU}) = 0 \tag{7.40}$$

$$\mathbf{E}'\mathbf{N} - \mathbf{E}'\mathbf{EU} = 0 \tag{7.41}$$

$$\mathbf{E}'\mathbf{EU} = \mathbf{E}'\mathbf{N} \tag{7.42}$$

$$\mathbf{U} = (\mathbf{E}'\mathbf{E})^{-1}\mathbf{E}'\mathbf{N} \tag{7.43}$$

$$\mathbf{EU} = \mathbf{E}(\mathbf{E}'\mathbf{E})^{-1}\mathbf{E}'\mathbf{N} = \mathbf{A}(\mathbf{A}'\mathbf{A})^{-1}\mathbf{A}'\mathbf{N}. \tag{7.44}$$

But $\mathbf{EU} = \mathbf{N}^*$; therefore

$$\mathbf{RN} = \mathbf{N}^*. \tag{7.45}$$

A similar analysis demonstrates that

$$\mathbf{I} - \mathbf{R} = \mathbf{B}. \tag{7.46}$$

I usually restrict matrix **R** operations to equation 7.20 and compute matrix **B** by equation 5.1.

The orthogonal projector is also closely related to the pseudoinverse or generalized inverse. Matrix **R** can be expressed as $\mathbf{A}(\mathbf{A}'\mathbf{A})^{-1}\mathbf{A}'$, whereas the pseudoinverse of **A** is $(\mathbf{A}'\mathbf{A})^{-1}\mathbf{A}'$. The pseudoinverse substitutes for the inverse in the solution of overdetermined simultaneous linear equations. Numerous textbooks and articles are available, including Nashed (1976), Boullion and Odell (1969, 1971), Brand (1962), Ficken (1967), Albert (1972), Ben-Israel and Greville (1975), Campbell and Meyer (1978), Amir-Moez and Newman (1970), Glencross (1979), James (1978), Planitz (1979), and Wong (1979). (See also Moore, 1920, and Penrose, 1956.)

*Fundamentals in Matrix **R** and Their Residuals*

Given only the definition of the orthogonal projector that produces Wyszecki's fundamentals, we may prove the other prong of the Wyszecki hypothesis: that the tristimulus values of any residual are always (0, 0, 0).

Here is the projection operator ejecting a fundamental, according to the first prong of the Wyszecki hypothesis:

$$\mathbf{A}(\mathbf{A}'\mathbf{A})^{-1}\mathbf{A}'\mathbf{N} = \mathbf{N}^*. \tag{7.47}$$

Then, by the Wyszecki hypothesis,

$$\mathbf{N} = \mathbf{N}^* + \mathbf{B} \text{ or } \mathbf{B} = \mathbf{N} - \mathbf{N}^*. \tag{7.48}$$

I shall prove that $\mathbf{A}'\mathbf{B} = \mathbf{Q} = \mathbf{0}$, where matrix **A** is any set of color mixture functions, matrix **B** is any metameric black, and matrix **Q** is the tristimulus values of the residual.

By definition,

$$\mathbf{A'B} = \mathbf{Q}. \tag{7.49}$$

Substituting,

$$\mathbf{A'[N - N^*]} = \mathbf{Q}. \tag{7.50}$$

Substituting from equation 7.29,

$$\mathbf{A'[N - A(A'A)^{-1}A'N]} = \mathbf{Q}. \tag{7.51}$$

Rearranging terms,

$$\mathbf{A'N - A'A(A'A)^{-1}A'N} = \mathbf{Q}. \tag{7.52}$$

But,

$$\mathbf{A'A(A'A)^{-1}} = \mathbf{I}. \tag{7.53}$$

Therefore,

$$\mathbf{A'N - IA'N} = \mathbf{A'N - A'N} = \mathbf{Q} = 0. \tag{7.54}$$

Given matrices **R** and $\mathbf{R_B}$, we may create metamers ad infinitum. We isolate the fundamental of any radiometric function with matrix **R**, and we isolate the residuals of other radiometric functions with $\mathbf{R_B}$. Then we add each residual to the fundamental, or even weighted combinations of the residuals to the fundamental, and each new radiometric function is a member of the same metameric suite. If a new radiometric function has negative values, the function will still be a theoretical member of the metameric suite, but the function is imaginary. These negative values often may be eliminated by multiplying the residual by a constant, positive or negative.

Figure 13 demonstrates the power of the Wyszecki hypothesis and matrix-**R** operations with respect to complementary stimuli and their metamers. Figure 13A is the radiometric function of an equal-energy stimulus, evoking a defined white sensation; the radiation is unit magnitude at every wavelength. Figure 13B is the fundamental of the equal-energy radiometric function with the characteristic bimodal distribution that is the sums of columns in matrix **R** (see Figure 7); the fundamental was computed by the basic equation, $\mathbf{RN} = \mathbf{N^*}$. I define the white sensation as that evoked by the function in *B* (or that in *A*). *C* depicts the radiometric function of Munsell GY 9/5 viewed under an equal-energy source, measured by Dorothy Nickerson (personal communication, 1955); and *D* is the fundamental of Munsell GY 9/5 obtained by matrix-**R** operations. *E* is the residual of the equal-energy radiometric function obtained by subtracting the fundamental in **B** from the radiometric function in **A.** *F* is the residual of *E*, added to the radiometric function in *A*. The functions in *A*, *B*, and *F* belong to the same metameric suite. The residual in *E*, added to the

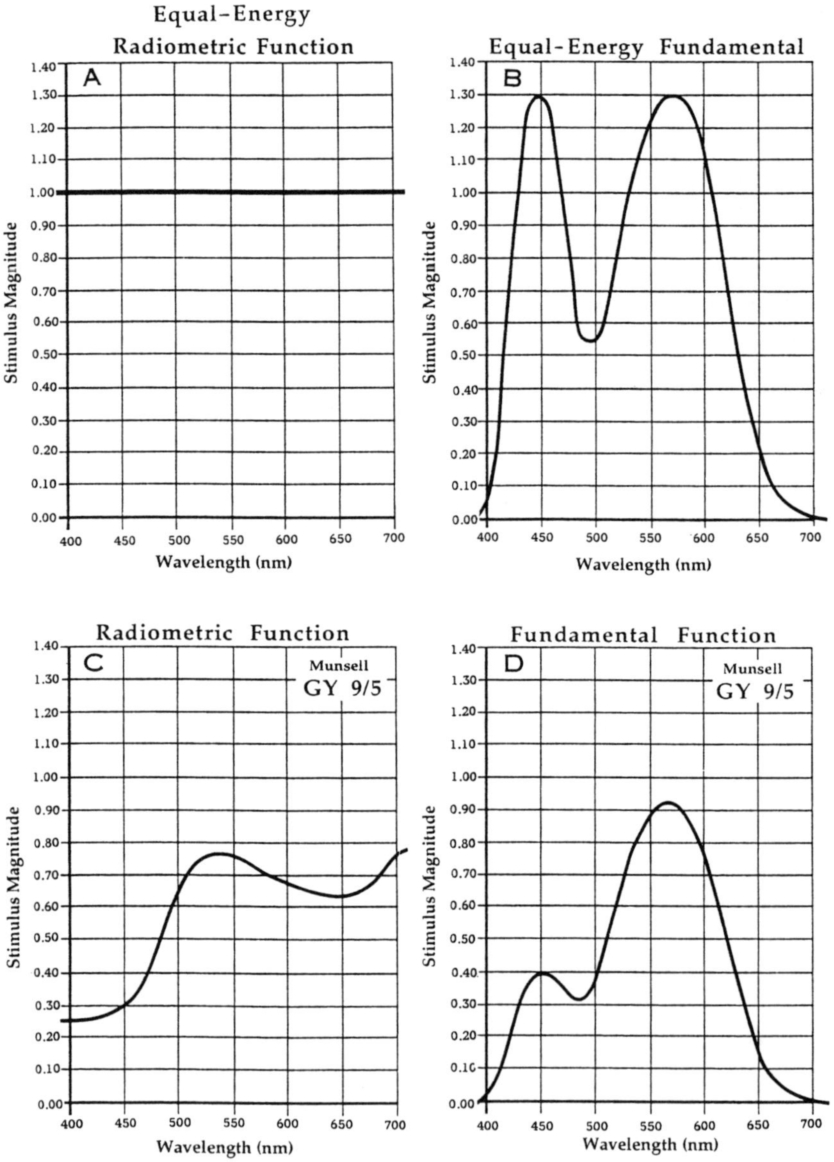

Figure 13. The Wyszecki hypothesis contributes to the understanding of complementarism in color mixture and color matching. *Continued on following pages.*

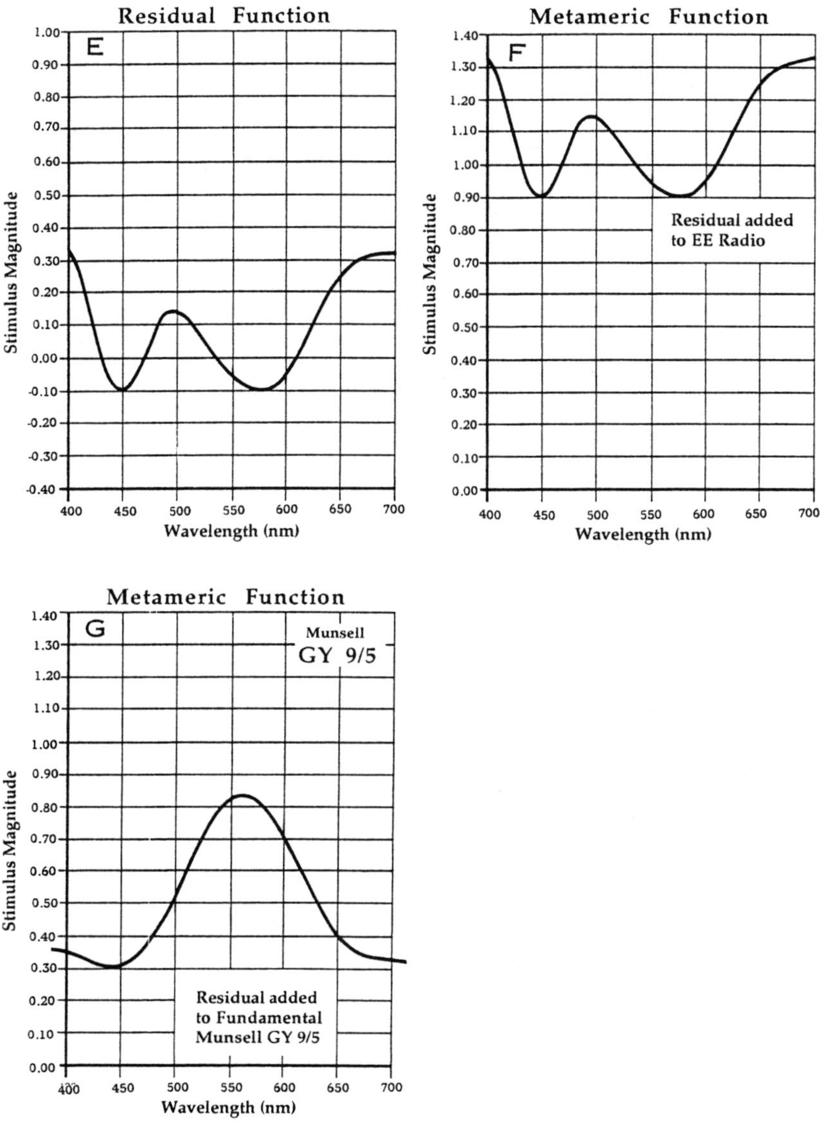

Figure 13. Continued.

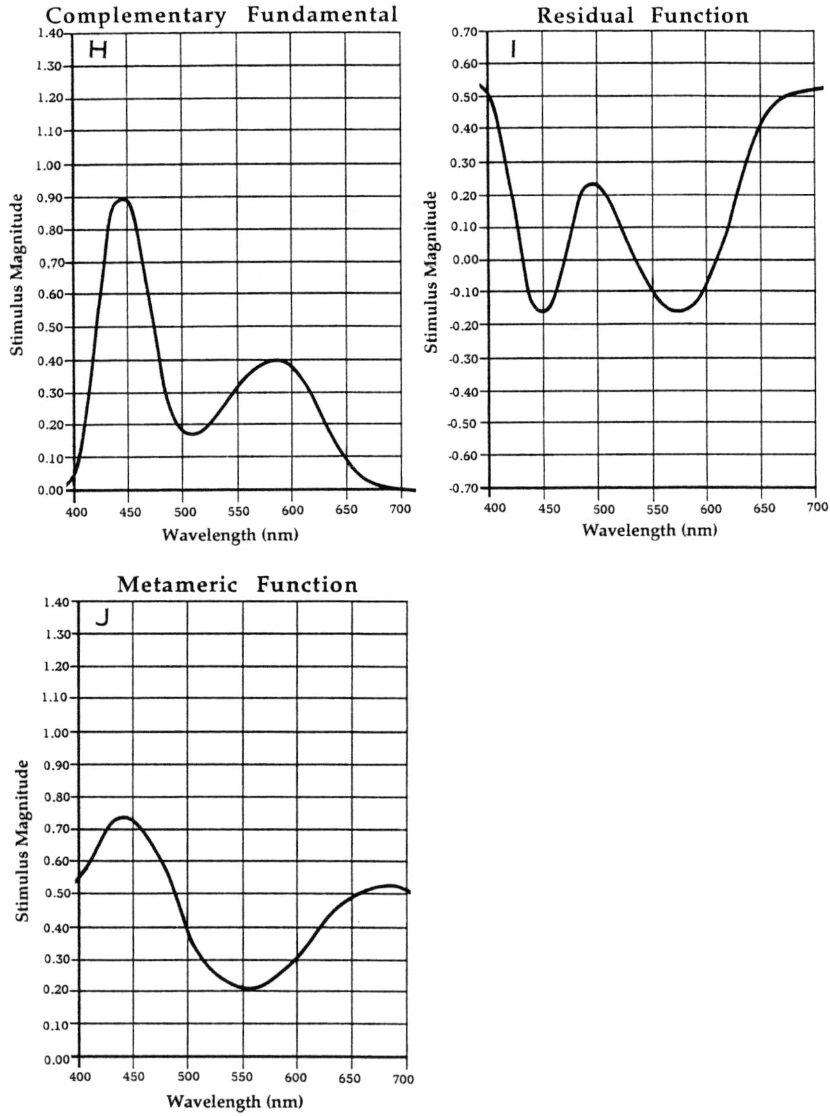

Figure 13. Continued.

fundamental of Munsell GY 9/5 in *D*, is a metamer to Munsell GY 9/5 in *G*. The functions in *C*, *D*, and *G* belong to the same metameric suite. *H* is the difference between the fundamental of the equal-energy stimulus in *B* and the fundamental of GY 9/5 in *D*; stated differently, the stimulus defined in *D* is complementary to the stimulus defined in *H*, because the compounded stimuli evoke a white sensation—that defined by the fundamental in *B*. *I* is a residual obtained from another Munsell chip. *J* is the sum of the residual function in *I*, and the fundamental in *H* and is a metamer to complementary function in *H*. The functions in *H* and *J* belong to the same metameric suite. Now, the functions in *C* and *D* individually are complementary to either of the functions in *H* and *J*, because each of the four pairs compounds to evoke a white sensation identical to that evoked by the functions in *A*, *B*, or *F*. These functions may be taken in any combination; for example, the function in *H* is complementary to the function in *C* because they compound to evoke the same white sensation as the function in *A*.

Additional discussion of complementary color stimuli appears later in this book.

*Richter's Metamers and Matrix-**R** Operations*

Colorimetrists generated metamers and residuals (but not fundamentals) before the advent of matrix **R** operations. The technique that received the most attention was developed by Manfred Richter (1958), described by Wyszecki and Stiles (1982), and used by Wyszecki (1958, 1962), Ohta and Wyszecki (1977), and Thornton (1973). Three of twelve metamers, computed by Wyszecki using the Richter technique, are illustrated in Figure 1.

Richter's method was laborious, and researchers had minimal control over the metamers generated; although the resulting metamers belonged to the same suite, the metamers were generated at a high level of randomness. Nevertheless, the method was the best available, and many researchers embraced the Richter technique in their studies of metamers and of residuals. (Residuals are produced when one metamer is subtracted from another metamer.) I shall demonstrate Richter's procedure in terms of matrix-**R** operations.

Although not made explicit by Richter, the maneuver uses the weak color-matching equation, 6.2.

Define:

Matrix **A**, comprising any three color matching functions , say x_t, y_t, z_t.

Matrix **E**, the fundamentals associated with the color-matching functions in matrix **A**.

Matrix $\mathbf{E} = (\mathbf{A}'\mathbf{A})^{-1}$, for reference purposes only (the matrix is not used in computations).

Matrix $\mathbf{N}_1$, a radiometric function representing one metamer in the metameric suite to be generated.

Matrix $\mathbf{N}_{a3}$, three arbitrary radiometric functions to be used in computing one function metameric to $\mathbf{N}_1$. The functions may be artificially or mathematically generated, as for example, gaussians. These are not the functions in matrix **E**.

In a color-matching experiment with the primaries in matrix **E**, the functions in matrix $\mathbf{N}_{a3}$ are matched ($\equiv$) when individually weighted by the tristimulus values given by Q3 (3×3, read by columns). As always,

$$\mathbf{A}'\mathbf{N}_{a3} = \mathbf{Q}_3. \tag{7.55}$$

In a color-matching experiment with the primaries in matrix **E**, the function in matrix $\mathbf{N}_1$ is matched ($\equiv$) when individually weighted by the tristimulus values given by Q1 (3×1). As always,

$$\mathbf{A}'\mathbf{N}_1 = \mathbf{Q}_1 \tag{7.56}$$

Define matrix **H** as a column matrix so that

$$\mathbf{Q}_3\mathbf{H} = \mathbf{Q}_1 \qquad \text{or} \qquad \mathbf{H} = \mathbf{Q}_3{}^{-1}\mathbf{Q}_1. \tag{7.57}$$

Premultiply by $\mathbf{A}(\mathbf{A}'\mathbf{A})^{-1}$, giving

$$\mathbf{N}_{a3}\mathbf{H} \equiv \mathbf{N}_1. \tag{7.58}$$

The left side of the equation matches the right side, and a metamer has been created. Note that equation 7.58 is also a weak color-matching equation, but the primaries are the radiometric functions in matrix $\mathbf{N}_{a3}$.

Another member of the metameric suite may be obtained with three different functions in $\mathbf{N}_{a3}$, now $\mathbf{N}_{b3}$. This may be continued indefinitely and any number of metamers computed. Why does the Richter method compute metamers in metameric suites? The answer is the character of the functions in the $\mathbf{N}_{a3}$ and similar matrices. These functions originally comprise different fundamentals and different residuals. The Richter process requires that the fundamentals of the $\mathbf{N}_{a3}$ matrix be transformed to the fundamental of the $\mathbf{N}_1$ matrix. The Richter process, however, does not control the residuals, and these are different for every triplet. This provides a random sum of residuals for the computed matrix that matches $\mathbf{N}_1$. If the functions in $\mathbf{N}_{a3}$ were fundamentals without residuals, new metamers would not be created; the resulting function would not only match $\mathbf{N}_1$ but would also be equal to $\mathbf{N}_1$.

This technique may be applied to the electronic transmisson and reception of color television. The color stimulus, entering the camera, is a radiometric function $\mathbf{N}_1$. The camera sensitivities may be any triplet of color-matching functions. The sensitivities are currently defined so that they evoke the sensations of blue, red, and green. This is not necessary. The sensitivities may evoke desaturated greens or purples or even three different whites. The only consideration is their deviance from color-matching functions.

This deviance can be measured. Premultiply the function by matrix **R**. The remainder, the residual, is a measure of error. When the residual is small, the error is small; this, not the sensation evoked by these sensitivities, is the principal consideration. Sensitivities with all positive values are preferred. Place these color-matching functions in matrix **A**.

Effectively, the sensitivities give the tristimulus values **Q** of the stimulus $\mathbf{N}_1$ computed by $\mathbf{Q} = \mathbf{A}'\mathbf{N}_1$. These are eventually transmitted as is or following a linear transformation.

At the receiver, the radiometric functions of the phosphors are gathered into matrix $\mathbf{N}_3$. (Theoretically these may be anything, but to avoid negative values, a wide separation produces a wide gamut of color sensations.) The 3×3 matrix $(\mathbf{A}_1'\mathbf{N}_3)$ is built into the receiver. The tristimulus values of matrix **Q** are premultiplied electronically by matrix $(\mathbf{A}_1'\mathbf{N}_3)^{-1}$. This gives matrix **H**, the weights or tristimulus values, specifying the analog voltages applied to the phosphors (after correction for gamma) to generate a metamer of $\mathbf{N}_1$ on the television screen. Standardization is unnecessary for the camera sensitivities or the television screen phosphors. Standardization is necessary only for the transmitted tristimulus values.

8

Mathematical Properties of Matrix **R**

Idempotency of Matrix R

In mathematical terms, matrix **R** is symmetrical, order *k*, rank 3, and idempotent—that is,

$$\mathbf{R} = \mathbf{RR}. \tag{8.1}$$

Because $\mathbf{R} = \mathbf{A}(\mathbf{A}'\mathbf{A})^{-1}\mathbf{A}'$, and substituting

$$\mathbf{A}(\mathbf{A}'\mathbf{A})^{-1}\mathbf{A}' = \mathbf{A}(\mathbf{A}'\mathbf{A})^{-1}\mathbf{A}'\mathbf{A}(\mathbf{A}'\mathbf{A})^{-1}\mathbf{A}'. \tag{8.2}$$

But $(\mathbf{A}'\mathbf{A})(\mathbf{A}'\mathbf{A})^{-1} = \mathbf{I}$, an identity matrix. Therefore,

$$\mathbf{A}(\mathbf{A}'\mathbf{A})^{-1}\mathbf{A}' = \mathbf{A}(\mathbf{A}'\mathbf{A})^{-1}\mathbf{I}\mathbf{A}' = \mathbf{R} = \mathbf{RR} \tag{8.3}$$

As a corollary, the sum of squares of each column in matrix **R** (including the square of the diagonal) equals the diagonal entry of that column.

Invariance of Matrix R under Linear Transformation

Matrix **R** is invariant; the identical matrix **R** may be computed from any matrix **A**, and because matrix **E** may always be computed from matrix **A**, matrix **R** may be computed from any matrix **E**.

Let **A** and $\mathbf{A}_a$ be two **A** matrices, one the transform of the other:

$$\mathbf{A}_a = \mathbf{AT}, \tag{8.4}$$

where **T** is an appropriate 3 × 3 transformation matrix (see below).

Let:

$$\mathbf{R}_a = \mathbf{A}_a(\mathbf{A}_a'\mathbf{A}_a)^{-1}\mathbf{A}_a'. \tag{8.5}$$

Then,

$$\mathbf{R}_a = \mathbf{AT}([\mathbf{AT}]'\mathbf{AT})^{-1}(\mathbf{AT})' \tag{8.6}$$

$$= \mathbf{A}(\mathbf{T}'\mathbf{A}'\mathbf{AT})^{-1}\mathbf{T}'\mathbf{A}'. \tag{8.7}$$

But,

$$(\mathbf{T}'\mathbf{A}'\mathbf{AT})^{-1} = \mathbf{T}^{-1}(\mathbf{A}'\mathbf{A})^{-1}(\mathbf{T}')^{-1}. \tag{8.8}$$

Therefore,

$$\mathbf{R}_a = \mathbf{ATT}^{-1}(\mathbf{A}'\mathbf{A})^{-1}(\mathbf{T}')^{-1}\mathbf{T}'\mathbf{A}' \tag{8.9}$$

$$= \mathbf{AI}(\mathbf{A}'\mathbf{A})^{-1}\mathbf{A}' \tag{8.10}$$

$$= \mathbf{A}(\mathbf{A}'\mathbf{A})^{-1}\mathbf{A}' \tag{8.11}$$

$$= \mathbf{R}. \tag{8.12}$$

Matrix **R** therefore is independent of the arbitrary primaries selected for the color-matching experiment. This invariance is appealing to researchers in the physical and psychophysical sciences. In an extraordinary essay, Max Born (1953) wrote: "In every physical theory, there is a rule which connects the projections of the same objects on different systems of reference, called a law of transformation, and all these transformations have the property of forming a group, i.e., the sequence of two transformations is a transformation of the same kind. *Invariants are quantities having the same value for any system of reference, hence they are independent of the transformation* [italics added]."

Matrix R, Fundamental Color Space, and F Matrices

Matrix **R**, of order *k* and rank 3, defines a configuration of vectors in the fundamental color space of three dimensions. The fundamental color space comprises vectors of invariant length separated by invariant angles. Each vector represents a unique color sensation and also the fundamental of the color stimulus evoking the color sensation. When the residuals are ejected from the radiometric functions, leaving only the fundamentals, color stimuli and color sensations are isomorphic; the same fundamental, scalar or vector, always evokes the identical color sensation. In fundamental color space, the vectors representing spectral color stimuli have novel internal harmonies and are aesthetically pleasant.

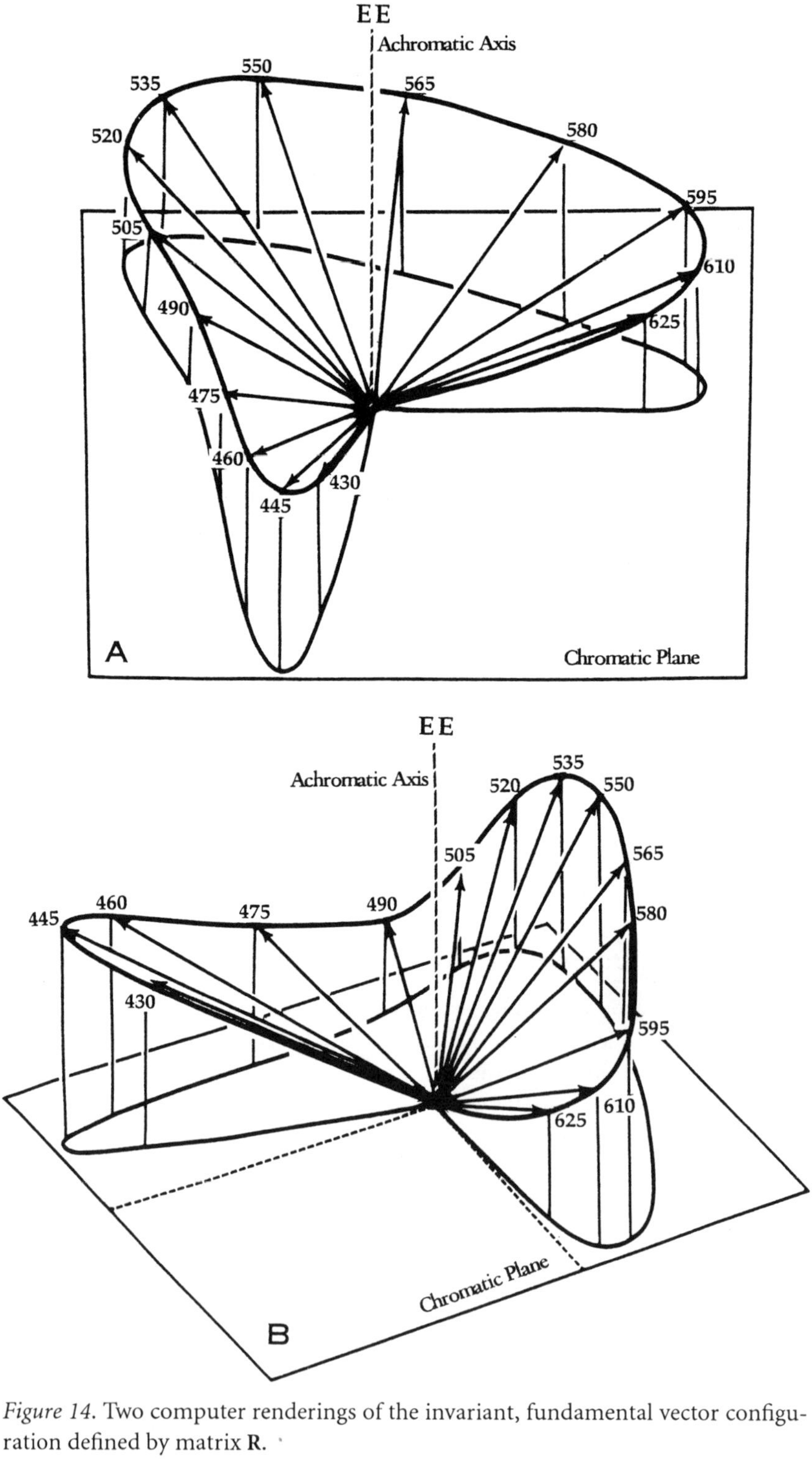

Figure 14. Two computer renderings of the invariant, fundamental vector configuration defined by matrix **R**.

This configuration of spectral vectors has been rendered by computer graphics and appears in Cohen (1988) and in Cohen and Kappauf (1985). Two new interpretations are shown in Figure 14.

The vectors represent monochromatic stimuli (and sensation) spanning the visible spectrum at 15 nm intervals. The vector lengths are the square roots of the diagonals in matrix **R**, and the vectors are separated by the appropriate scalar products in matrix **R**. The monochromatic vectors lie essentially on two "wings" (better observed in Figure 15). The fundamental vectors compound by vector addition, exactly as force vectors compound by vector addition. Thus, real heterochromatic color stimuli (and color sensations) lie within the configuration. Imaginary color stimuli (and sensations) lie outside the configuration. The fundamental vector configuration is independent of all triplets of reference axes. However, this display introduces an "achromatic" axis coincident with the "white" equal-energy vector at the centroid, and two "chromatic" mutually orthogonal axes on a plane orthogonal to the "achromatic" axes. The scalar fundamental of every vector is the locus of scalar products between the given vector and all other monochromatic vectors on the envelope.

Figure 15 illustrates the rotation of the fundamental color space, comprising two "butterfly" wings, where the spectral vectors hug two intersecting planes. Figure 15A is a side view of the vector configuration; this view has become the logo of matrix-**R** operations. In *B*, the configuration has been rotated about a vertical axis; the internal **V** structure becomes apparent. *C* shows rotation of more than 180 degrees about a horizontal axis, and the vector structure becomes more evident. *D* continues the rotation about the horizontal axis until the configuration is viewed from directly above. *D* also depicts the projections of the monochromatic vectors onto the "chromatic" planes; the angles between the monochromatic vector projections are critical to my arguments concerning complementarism. Note that the spectral vectors cluster at the convolution points on the spectral envelope; we shall inspect this clustering later.

Again we note that the scalar fundamental is represented in the immutable, fundamental vector configuration; the scalar fundamental is the locus of scalar products between the given color vector and the vector representing the spectral, monochromatic stimuli that lie on the envelope of the configuration. This is always true, even for the monochromatic vectors themselves. Scalar fundamentals do not depend on an arbitrary reference frame introduced into the vector configuration.

With respect to the vector fundamentals, however, we may select three independent reference or basis vectors. The choice of reference axes is arbitrary. I may choose monochromatic or heterochromatic vectors, real or imaginary,

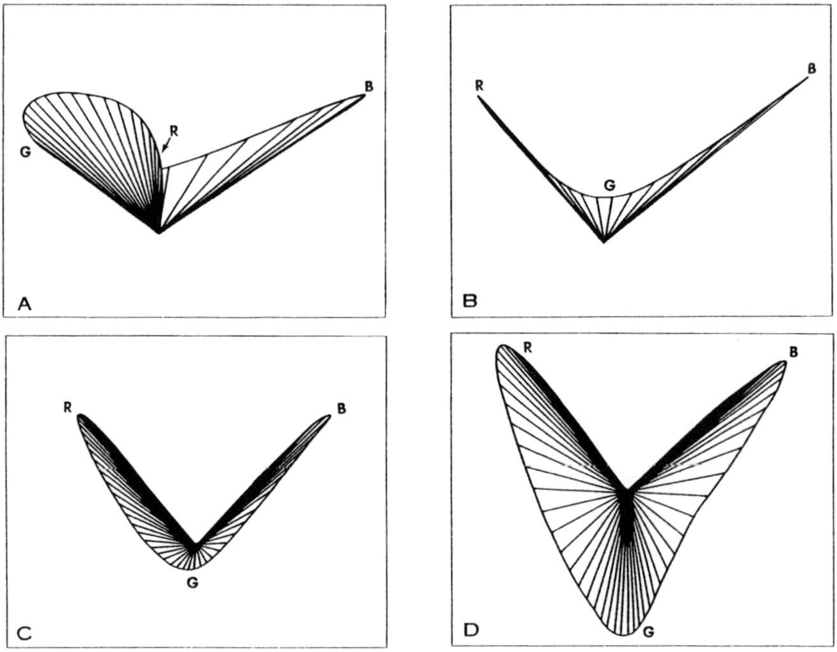

Figure 15. Four additional computer renderings of the monochromatic vectors in the fundamental color space (see Figure 14).

orthogonal or oblique, or any other combination. Although the choice is arbitrary, the vector configuration itself is absolutely immutable and does not change with a different selection of axes.

Although the choice of reference axes is vast, the mathematics is considerably easier when the reference axes are orthogonal, and even easier when the axes are normalized. Therefore, I seek an **F** matrix with three columns representing reference axes both orthogonal and normalized, termed *orthonormal.* Any given vector (representing a color sensation and its fundamental) may be specified by its projections onto the three orthonormal axes. The projections of the spectral vectors onto the three orthonormal axes comprise an **F** matrix; each spectral color stimulus has three projections, and the triplet of projections are the entries in one row of matrix **F**. The configuration harbors an infinite number of orthonormal triplets. In matrix **R** colorimetry, the columns of an **F** matrix are termed *tricolor orthonormal functions* or simply *tricolor functions,* in contrast to the color-matching functions in matrix **A** and the fundamentals in matrix **E**.

When a $(k \times 3)$ **F** matrix is orthonormal,

$$\mathbf{FF}' = \mathbf{R}. \tag{8.13}$$

An infinite number of **F** matrices, one the orthogonal transformation of another, specify the same matrix **R**. Stated differently, matrix **R** is invariant with respect to the orthogonal linear transformation of **F** matrices. Given an **A**, **E**, or **R** matrix, an infinite number of **F** matrices may always be computed, but to satisfy equation 8.13, all reproduce the same **R** matrix. This assertion will be proved presently.

I examine two techniques for extracting an **F** matrix from the fundamental color space defined by matrix **R**, give explicit procedures for deriving invariant vector lengths and invariant intervening angles, and detail the trace (sum of diagonal entries) in matrix **R**.

The so-called Gram-Schmidt orthogonalization method is a common computational device and appears in most textbooks on linear algebra. The procedure has an intriguing, largely unknown history. The technique was invented separately by Jorgen Pederson Gram and Erhardt Schmidt, distinguished mathematicians in the nineteenth and twentieth centuries; from Gram's work on matrix "factoring," the word *gramian* entered the mathematical language. The orthogonalization procedure was the subject of Gram's Danish doctoral dissertation in 1879, later published in German (1883). Independently, the orthogonalization procedure was also the subject of Schmidt's doctoral dissertation in German at the University of Göttingen (1907). Schmidt used materials unavailable to Gram, but both suspected that the orthogonalization procedure was closely allied to the least-squares technique utilizing orthogonal projectors. Schmidt graciously acknowledged Gram's contribution, and the technique bears both names. (See Longley, 1984.) Appendix A contains a BASIC program for the Gram-Schmidt procedure.

Nevertheless, Gram-Schmidt orthogonalization depends on difficult concepts of vector components in different directions and the subsequent subtraction of these components; the columns of matrix **F** are calculated successively. However, Gram-Schmidt orthogonalization is duplicated precisely, with far greater efficiency, by matrix-**R** operations where the columns of matrix **F** emerge simultaneously. This newer process is developed herewith, so that direct comparison may be made to Gram-Schmidt.

Orthonormal Functions Derived from Matrix R

To obtain orthonormal reference axes in fundamental color space, I begin with three independent fundamentals in color space. (This procedure is illustrated in

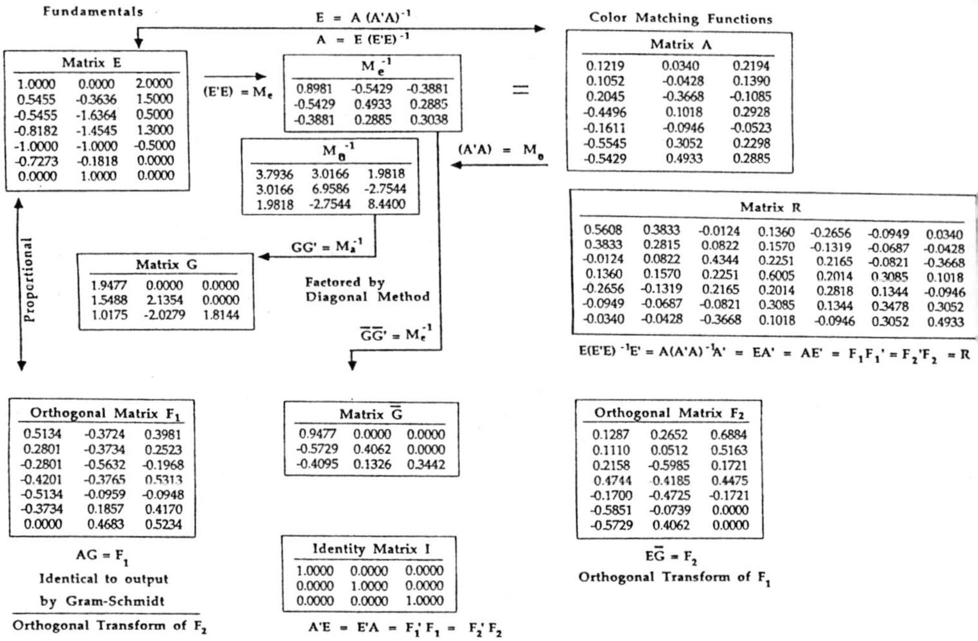

Figure 16. Flow diagram of matrix **R** computational procedures to orthogonalize any $k \times 3$ matrix of fundamentals or color-matching functions.

Figure 16). The procedure guarantees results identical to those obtained with the Gram-Schmidt orthogonalization procedure, where the first column of the orthonormal functions is proportional to the first column of the input matrix. Each fundamental is a $k \times 1$ matrix, and the three fundamentals may be gathered into a single $k \times 3$ matrix **E**, where the columns are $\mathbf{E}_1$, $\mathbf{E}_2$, and $\mathbf{E}_3$, respectively. The columns of the **E** matrix are not orthogonal; $\mathbf{E}'\mathbf{E}$ does not give a 3×3 identity matrix. Given the **E** matrix, the goal is to derive three new $k \times 1$ matrices $\mathbf{F}_1$, $\mathbf{F}_2$, and $\mathbf{F}_3$; these new functions may be gathered into an **F** matrix so that $\mathbf{F}'\mathbf{F} = \mathbf{I}$. Each of these six fundamentals is represented by a vector in fundamental color space. The angles between the **E** vectors are not orthogonal. The angles between the **F** vectors *are* orthogonal.

We may, in fact, begin with any three independent fundamental vectors designated by matrix **E** and derive an **F** matrix. Almost invariably, we are interested in the precise placement of the first reference axis; the first column of matrix **E** is identical to the first column of matrix **F**, except for normalization so that the sum of squares equals unity in the **F** matrix.

The three column vectors in **E** (representing the color fundamentals) and the three column vectors in **F** (also representing three color fundamentals) are all vectors in fundamental color space. The vectors in **E** are at oblique angles to each other and not normalized; the vectors in **F** are orthogonal to each other and normalized. We have **E**, and we want **F**—a particular **F** where the first column is proportional to the first column in matrix **E**. Suppose, as is often the case, that we need the fundamental of an equal-energy stimulus as the first column of matrix **E**. That fundamental is obtained by matrix-**R** operations, $\mathbf{RN} = \mathbf{N}^*$, where **N** is a "flat" radiometric function with equal irradiance at every wavelength. This particular bimodal fundamental is the sum of the columns (or rows) in matrix **R**, and is illustrated in Figure 7B and Figure 13B. This fundamental therefore is $k \times 1$ matrix $\mathbf{F}_1$. We still need $\mathbf{F}_2$ and $\mathbf{F}_3$.

We obtain an orthonormal triplet of vectors using the extraordinary versatility of matrix **R**. Given equation 7.6, $\mathbf{A}(\mathbf{A}'\mathbf{A})^{-1}\mathbf{A}' = \mathbf{A}(\mathbf{M}_a)^{-1}\mathbf{A}' = \mathbf{R}$. Matrix $(\mathbf{M}_a)^{-1}$ is gramian as a condition deriving from its computation and therefore may be resolved into three components, columns of matrix **G**, such that

$$\mathbf{GG}' = (\mathbf{M}_a)^{-1}. \tag{8.14}$$

Thus,

$$\mathbf{AGG}'\mathbf{A}' = \mathbf{R} \tag{8.15}$$

and

$$\mathbf{AG} = \mathbf{F}. \tag{8.16}$$

The important element in equation 7.6 is $(\mathbf{M}_a)^{-1}$, equal to $\mathbf{M}_e$, the scalar products between three arbitrary fundamentals in fundamental color space. An interesting problem arises. We can begin with matrix **E** and use the equation $\mathbf{E}(\mathbf{E}'\mathbf{E})^{-1}\mathbf{E}' = \mathbf{R}$. Remembering that $(\mathbf{E}'\mathbf{E})^{-1} = \mathbf{M}_a$ and that $(\mathbf{A}'\mathbf{A})^{-1} = \mathbf{M}_e$, the following method will operate on matrix **A** and produce an **F** matrix whose first column is proportional to the first column in matrix **E**. On the other hand, the following method will operate on matrix **E** and produce an **F** matrix whose first column is proportional to the first column in matrix **E**.

Therefore, to use the following procedure most effectively, gather three independent fundamentals together as an **E** matrix. The first column will designate the vector representing the first reference axis; often this is the fundamental of an equal-energy stimulus. Then use the now-familiar $\mathbf{A} = \mathbf{E}(\mathbf{E}'\mathbf{E})^{-1}$ to transform this **E** matrix to its corresponding **A** matrix. The emergent **A** matrix may give a peculiar appearance, showing almost no resemblance to the **E** matrix from which it was derived. Nevertheless, this **A** matrix is to be used as the input. Then, the first column in the resulting **F** matrix will be proportional to the first column in the **E** matrix, *not* the first column in the derived **A** matrix.

This resolution of $(\mathbf{M}_a)^{-1}$ into **G** matrices is often termed *factoring*, and an infinite number of **G** matrices fulfill the conditions of equation 8.14. The particular **G** matrix depends on the factoring method, and each **G** matrix gives a different **F** matrix. Probably the simplest factoring technique is the so-called diagonal method, described in the literature by many authors (for example, by Fox, 1963).

In the diagonal method, matrix **G** has the form:

$$\begin{bmatrix} g_{11} & 0 & 0 \\ g_{21} & g_{22} & 0 \\ g_{31} & g_{32} & g_{33} \end{bmatrix} = \mathbf{G}. \tag{8.17}$$

This **G** matrix requires the computation of only six of nine elements in symmetrical 3 × 3 matrix $(\mathbf{M}_a)^{-1} = \mathbf{M}_e$.

The 3 × 3 matrix **G** is the transformation matrix that carries matrix **E** into one matrix **F**. When computed in this manner, the first column of matrix **F** will be proportional to the first column of matrix **E**, and the sums of squares in the first column of matrix **F** equals unity. However, the remaining two columns of matrix **E** in general will not correspond to the remaining two columns in matrix **F**, for the reason that the second two columns in matrix **E** are not orthogonal to the first column or to each other, while the second two columns in matrix **F** are orthogonal to the first column and also orthogonal to each other.

I demonstrate the efficacy of this matrix-**R** technique for orthogonalization; this is best accomplished by matrix equations, not geometry. As indicated above, $\mathbf{F}'\mathbf{F} = \mathbf{I}$, a statement that every **F** matrix fulfills the mathematical tensor quantity known as the Kronecker delta. In matrix **F**, the sum of squares in each of the three columns must equal unity, and the "cross-products" between columns must equal zero.

Given equation 7.6,

$$\mathbf{A}(\mathbf{A}'\mathbf{A})^{-1}\mathbf{A}' = \mathbf{R}.$$

Let

$$\mathbf{F} = \mathbf{AG}. \tag{8.18}$$

Then,

$$\mathbf{F}'\mathbf{F} = \mathbf{G}'\mathbf{A}'\mathbf{AG}. \tag{8.19}$$

Postmultiply by **G**′,

$$\mathbf{F'FG'} = \mathbf{G'A'AGG'}. \tag{8.20}$$

However,

$$\mathbf{GG'} = (\mathbf{A'A})^{-1}. \tag{8.21}$$

Hence,

$$\mathbf{F'FG'} = \mathbf{G'A'A}(\mathbf{A'A})^{-1} = \mathbf{G'}. \tag{8.22}$$

Postmultiply by $(\mathbf{G}')^{-1}$,

$$\mathbf{F'FG'}(\mathbf{G'})^{-1} = \mathbf{G'}(\mathbf{G'})^{-1} = \mathbf{F'F} = \mathbf{I}. \tag{8.23}$$

Also,

$$\mathbf{FF'} = \mathbf{R}. \tag{8.24}$$

This method of deriving an **F** matrix, using matrix **R** operations, is markedly efficient, but is not discussed in textbooks on linear algebra, as far as I am aware. The reason, I believe, is that an **A** matrix (and not an **E** matrix) was generally used as the input. This provided one **F** matrix, to be sure, but the first column (axis) was not proportional to the input **A**. When the primary input is an **E** matrix, as should be the case because we are dealing with fundamentals, and the corresponding **A** matrix used as the secondary input, then the result is exactly equivalent to the almost universal Gram-Schmidt procedures. Appendix B contains a BASIC program for calculation of an **F** matrix by matrix-**R** operations.

The Trace Reconsidered

We can now understand the mathematical theorem stating that the trace (sum of the diagonals) of a matrix is precisely equal to three (see Figure 7C). This is characteristic of idempotent matrices in general, where the trace equals the dimensionality—three in fundamental color space. Given that $\mathbf{FF'} = \mathbf{R}$, this means that the sum of squares of each column is equal to unity, and since there are three columns, the sum of the square of every entry equals three. Each diagonal entry in matrix **R** equals the sum of squares of the entries in the corresponding row (or column). Thus, the sum of the diagonals must also equal three, which is the dimensionality of matrix **R**. This theorem is invariably sustained regardless of the number of rows in the **E** matrix used to compute matrix **R**.

Computation of Vector Lengths and Angles Defined by Matrices F and R

Recall that the entries in matrix **R** are scalar products between spectral vectors and also that $\mathbf{FF}' = \mathbf{R}$, where the entries in each row of matrix **F** are the coordinates of a spectral vector with respect to the three arbitrary orthogonal reference axes. However, every fundamental, real and within the envelope or imaginary and outside the envelope, also has coordinates with respect to the reference axes of matrix **F**. Given a matrix **N** or $\mathbf{N}^*$, the radiometric function or fundamental of any color stimulus, then

$$\mathbf{F}'\mathbf{N} = \mathbf{F}'\mathbf{N}^* = \mathbf{SS}, \tag{8.25}$$

where **SS** is a 3×1 matrix of the coordinates of the stimulus. Matrix $\mathbf{SS}'$ is then an appropriate additional row in matrix **F**.

Suppose that we choose to find the vector lengths and intervening angles of the fundamentals specifying the CIE 1964 observer R,G,B and X,Y,Z. Remember that the CIE presents these data as color-matching functions. Immediately, we operate on these data triplets to obtain matrices $\mathbf{E}_1$ and $\mathbf{E}_2$ of the respective fundamentals, using the now-familiar equation $\mathbf{E} = \mathbf{A}(\mathbf{A}'\mathbf{A})^{-1}$. Then $\mathbf{F}'\mathbf{E}_1 = \mathbf{SS}_1$ and $\mathbf{F}'\mathbf{E}_2 = \mathbf{SS}_2$, where the columns of $\mathbf{SS}_1$ and $\mathbf{SS}_2$ are the coordinates of R,G,B and X,Y,Z fundamentals with respect to the reference axes defined by matrix **F**. Taking $\mathbf{SS}'_1$ and $\mathbf{SS}'_2$ together, we have six rows in a miniature **F** matrix termed $\mathbf{F}_c$. Then $\mathbf{F}_c\mathbf{F}'_c = \mathbf{R}_c$, where the entries in a 6×6 symmetrical matrix $\mathbf{R}_c$ are scalar products analogous to the scalar products in matrix **R**.

I may compute the lengths and angular separations between any two vectors (with coordinates in matrix $\mathbf{F}_c$), $\vec{C}_1$, and $\vec{C}_2$, from the scalar product between these vectors (with an appropriate entry in matrix $\mathbf{R}_c$).

Recall that the scalar product is defined

$$\vec{C}_1 \cdot \vec{C}_2 = |\vec{C}_1| \cdot |\vec{C}_2| \cdot \cos\phi. \tag{8.26}$$

The respective vector lengths are the square roots of the diagonals in matrix $\mathbf{R}_c$ or, what is the same thing, the square roots of the sum of the respective row entries in matrix $\mathbf{R}_c$ (termed the *norm*, the Pythagorean theorem in three dimensions). To obtain the angle between two vectors, multiply the two vector lengths (square roots of the diagonals), and divide by the scalar product between the two vectors, the appropriate entry in matrix $\mathbf{R}_c$. This gives the cosine of the angle, sometimes called the *direction cosine*. The angle, then, is the arc cosine of the cosine, and may be obtained from an inexpensive hand calculator. Appendix C contains a BASIC program for these computations. Strangely, the arc cosine function is not found in many computer application programs, including BASIC, but it may be computed from the arc tangent function by this relationship:

$$\text{arc cosine } \phi, = 2 \arctan[(1 - q)/(1 + q)]^{1/2}. \tag{8.27}$$

The example has used six fundamentals derived from two triplets of CIE color-matching functions. The results are presented later, where the representation of the CIE primaries in fundamental color space is detailed. The procedure is applicable to any number of vectors.

The more common colorimetric problem, however, is the calculation of the angles between three vectors in the fundamental color space. These angles are most easily computed from $\mathbf{M}_e$ matrices, not from $\mathbf{M}_a$ matrices. (See equations 7.6 and 7.7.)

The $\mathbf{M}_e$ matrix comprises the scalar products between the axes. Given an **A** matrix or an **E** matrix, the $\mathbf{M}_e$ matrix is easily computed, and the calculations are analogous. The diagonals of the $\mathbf{M}_e$ matrix (not the $\mathbf{M}_a$ matrix), like the diagonals of the **R** matrix, are the squares of the vector lengths, and the side entries, like the side entries in matrix **R**, are the scalar products. To compute the angles, first compute the square roots of the diagonals, which are the vector lengths. Then multiply pairs of vector lengths, and divide into the appropriate scalar product to obtain the direction cosine. Then the angle is the arc cosine of the direction cosine. I repeat: the correct answer is obtained from matrix $\mathbf{M}_e$, not from $\mathbf{M}_a$. Researchers have used matrix $\mathbf{M}_a$ and have obtained absurd angles between the CIE primaries; this will be detailed.

9

Scalar Products: The Entries in Matrix R

Definition of Scalar Products

The scalar product was defined by Hermann Günther Grassmann, the same nineteenth-century researcher who wrote the insightful article usually considered to be the basis of modern colorimetry. Grassmann, however, never related his scalar products to the vector representation of color stimuli and color sensations discussed in this book; scalar products, however, permeate modern physics. The vector product, sometimes called the outer or cross-product, was also defined by Grassmann but has not yet found a place in modern colorimetry.

The scalar product between vectors $\vec{C}_1$ and $\vec{C}_2$ was defined previously, and is repeated here because of its saliency:

$$\vec{C}_1 \cdot \vec{C}_2 = |\vec{C}_1| \cdot |\vec{C}_2| \cdot \cos\phi, \tag{9.1}$$

where $|\vec{C}_1|$ and $|\vec{C}_2|$ are the respective vector lengths and ϕ is the intervening angle.

I introduce an arbitrary orthogonal reference frame into the vector configuration defined by matrix **R** and illustrated in Figure 17A. Arbitrarily, I designate the axes of this reference frame as b, g, and r, but this is merely a convenience; the axes need not represent blue, green, or red color stimuli or color sensations. The coordinates of $\vec{C}_1$ and $\vec{C}_2$ are (b_1, g_1, r_1) and (b_2, g_2, r_2) respectively; each triplet of coordinates is a separate row in matrix **F**.

Figure 17A shows vectors $\vec{C}_1$ and $\vec{C}_2$, the negative vector $-\vec{C}_1$, and the vector addition $(\vec{C}_2 - \vec{C}_1)$. Observe that the length of the vector $(\vec{C}_2 - \vec{C}_1)$ is $|\vec{C}_2 - \vec{C}_1|$ and is the distance between the arrow heads of vectors $\vec{C}_2$ and $\vec{C}_1$.

By the law of cosines,

$$2|\vec{C}_1||\vec{C}_2| \cos\phi = |\vec{C}_1|^2 + |\vec{C}_2|^2 - |\vec{C}_2 - \vec{C}_1|^2. \tag{9.2}$$

By virtue of the orthogonal frame and the Pythagorean theorem,

$$|\vec{C}_1| = [(b_1)^2 + (g_1)^2 + (r_1)^2]^{.5} \tag{9.3}$$

$$|\vec{C}_2| = [(b_2)^2 + (g_2)^2 + (r_2)^2]^{.5}. \tag{9.4}$$

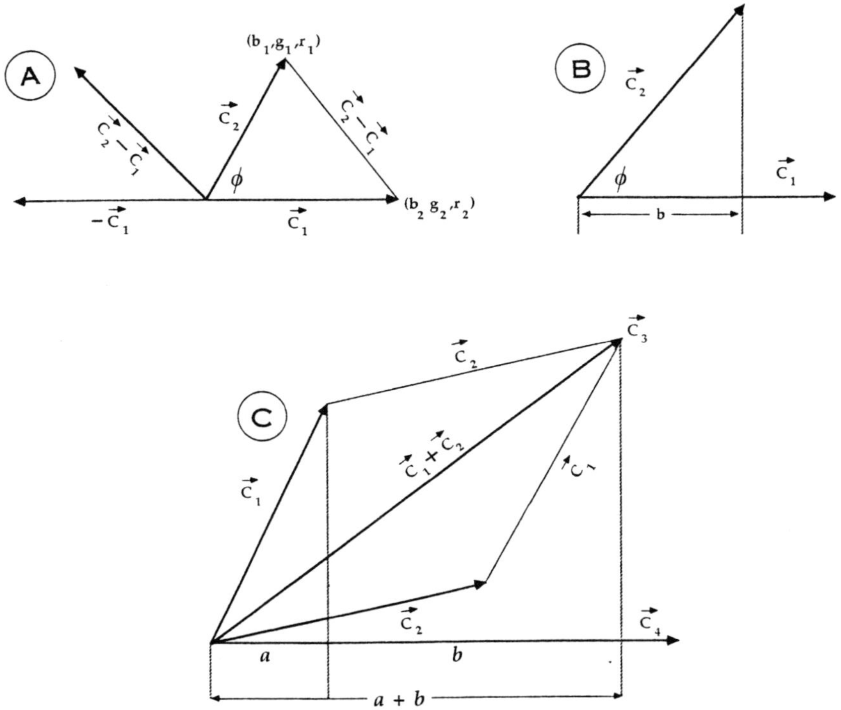

Figure 17. Vector diagrams employed in proofs of three important theorems concerning the fundamental color space; see text. *A* is referenced in proof of the theorem: Entries in matrix **R** are scalar products between vectors representing monochromatic fundamental color stimuli. *B* is referenced in proof of the theorem: The scalar product between two vectors (each representing the fundamental of a color stimulus) may be interpreted as the product of the projection of the second vector onto the first vector and the length of the first vector. *C* is referenced in proof of the theorem: Scalar products (entries) in matrix **R** are scalarly additive by rows or columns.

By the distance equation in analytic geometry,

$$|\vec{C}_2| - |\vec{C}_1| = [(b_1 - b_2)^2 + (g_1 - g_2)^2 + (r_1 - r_2)^2]^{0.5}. \qquad (9.5)$$

Substituting in equation 9.2,

$$2|\vec{C}_1||\vec{C}_2|\cos\phi = [(b_1)^2 + (g_1)^2 + (r_1)^2] + [(b_2)^2 + (g_2)^2 + (r_2)^2] - [(b_1 - b_2)^2 + (g_1 - g_2)^2 + (r_1 - r_2)^2]. \qquad (9.6)$$

But,

$$-[(b_1 - b_2)^2 + (g_1 - g_2)^2 + (r_1 - r_2)^2] = -(b_1)^2 + (2b_1b_2) - (b_2)^2 - (g_1)^2 + (2g_1g_2) - (g_2)^2 - (r_1)^2 + (2r_1r_2) - (r_2)^2. \qquad (9.7)$$

Therefore,

$$2|\vec{C}_1||\vec{C}_2|\cos\phi = 2(b_1b_2 + g_1g_2 + r_1r_2)$$

or

$$|\vec{C}_1||\vec{C}_2|\cos\phi = b_1b_2 + g_1g_2 + r_1r_2 = \vec{C}_1 \cdot \vec{C}_2. \qquad (9.8)$$

Recall that **FF**$'$ = **R**. The left segment of equation 9.8 is the definition of the scalar product. The middle segment is the entry in matrix **R** between vectors $\vec{C}_1$ and $\vec{C}_2$. The right segment is the "cross-product" between the entries in two rows of any **F** matrix; because **FF**$'$ = **R**, the entries in matrix **R** are scalar products. This is a powerful technique for obtaining the scalar product between two vectors with known coordinates in an orthonormal frame.

Geometry of Scalar Products

Matrix-**R** operations are totally dependent on Grassmann's concept of the scalar product. Here we examine three aspects of its geometry.

Figure 17B depicts two fundamental vectors, $\vec{C}_1$ and $\vec{C}_2$. Vector $\vec{C}_2$ is projected onto $\vec{C}_1$. By the cosine definition, b = $|\vec{C}_2|\cos\phi$. Multiplying by $|\vec{C}_1|$, we have

$$|\vec{C}_1|\mathrm{b} = |\vec{C}_1||\vec{C}_2|\cos\phi. \qquad (9.9)$$

The right side of this equation is the definition of the scalar product. Therefore, the scalar product may be expressed using a projection: the length of the vector representing the first fundamental multiplied by the projection of the vector representing the second fundamental onto the vector representing the first fundamental.

The scalar products of fundamentals add distributively. Figure 17C illustrates vectors for two fundamentals, $\vec{C}_1$ and $\vec{C}_2$, their vector sum $\vec{C}_3 = (\vec{C}_1 + \vec{C}_2)$, and another vector representing fundamental $\vec{C}_4$. In Figure 17C, the projection of $\vec{C}_1$ onto $\vec{C}_4$ plus the projection of $\vec{C}_2$ onto $\vec{C}_4$ equals the projection of $\vec{C}_3 = \vec{C}_1 + \vec{C}_2$ onto $\vec{C}_4$. Referring to the theorem developed in the preceding paragraph, the scalar product between $\vec{C}_1$ and $\vec{C}_4$ plus the scalar product between $\vec{C}_2$ and $\vec{C}_4$ equals the scalar product of $\vec{C}_3$ (vector sum of $\vec{C}_1$ and $\vec{C}_2$) and $\vec{C}_4$; more succinctly,

$$\vec{C}_1 \cdot \vec{C}_4 + \vec{C}_2 \cdot \vec{C}_4 = \vec{C}_3 \cdot \vec{C}_4, \qquad \text{where} \qquad \vec{C}_1 + \vec{C}_2 = \vec{C}_3.$$

This gives the important theorem: Fundamentals, as conceived by Wyszecki and as represented in matrix **R**, add by scalar addition. Each row (column) of matrix **R** is the fundamental of a monochromatic stimulus; each fundamental is also represented by a fixed vector in Euclidean fundamental color space. Scalar products between $\vec{C}_1$ and $\vec{C}_4$ and between $\vec{C}_2$ and $\vec{C}_4$ are entries for $\vec{C}_1$ and $\vec{C}_2$ in column labeled C_4, matrix **R**. When vectors $\vec{C}_1$ and $\vec{C}_2$ are combined by vector addition (additive color mixture) to give $\vec{C}_3$, the scalar product between $\vec{C}_3$ and $\vec{C}_4$ is the scalar sum of the entries for $\vec{C}_1$ and $\vec{C}_2$ in column labeled C_4, matrix **R**. Generalizing, scalar fundamentals, comprising scalar products, may be combined by scalar addition.

The vector ordering in the scalar product is commutative, as illustrated in Figure 17C; $\vec{C}_1 + \vec{C}_2 = \vec{C}_2 + \vec{C}_1$. Thus, the symmetrical entries in matrix **R** are equal.

Therefore, the scalar product between orthogonal vectors is 0 ($\cos 90^\circ = 0$), and the scalar product between coincident unit vectors is 1.00 ($\cos 0^\circ = 1.00$). The scalar product of a vector with itself is the square of the vector length:

$$\vec{C}_1 \cdot \vec{C}_1 = |\vec{C}_1||\vec{C}_1|1 = |\vec{C}_1|^2. \tag{9.10}$$

Thus, the squared vector lengths of the monochromatic vectors, on the envelope of the vector configuration, are the diagonal entries in matrix **R**.

10

Matrices **A** and **E** and **F**: Interactions and Connections

Matrix **A** is the matrix of color-matching functions. Matrix **E** is the matrix of primary fundamentals used in the experiment that generated matrix **A**. Matrix **F** is the matrix that projects spectral vectors into three orthonormal axes.

The fundamental of every color stimulus, monochromatic or heterochromatic, is absolutely invariant. The same fundamental forever belongs to the same color stimulus. The color-matching functions, however, are highly variable. (See the demonstration in Cohen, 1988.)

Suppose that we use one triplet of primaries to generate one triplet of color-matching functions. Given the emerging matrix **A**, we can immediately compute the fundamentals of the primaries, matrix **E**. Now we change one primary, and again generate a triplet of color-matching functions, another matrix $\mathbf{A}_1$. In general, no column of matrix **A** will be the same as any other column in matrix $\mathbf{A}_1$, although only one primary was changed. As before, we compute the fundamentals of the new primaries, $\mathbf{E}_1$. The two unchanged primaries are represented by columns in **E** and will have corresponding columns in $\mathbf{E}_1$. These corresponding columns in matrix **E** and $\mathbf{E}_1$ will be identical, although the color-matching functions are very different. The third changed primary, represented by the third column in matrix **A**, will correspond to a very different column in matrix **E**. This constancy of fundamentals with respect to their stimuli, and also this dependency of color-matching functions on the choice of the other two primaries as indicated, was illustrated in Cohen (1988). MacAdam (1942) had previously noted the dependency of color-matching functions but not the constancy and invariance of fundamentals of the primaries.

These constancies can perhaps be demonstrated more appropriately with **M** matrices. Figure 18 demonstrates miniature matrices, **A** and $\mathbf{A}_1$. Matrix **A** has primaries at row vectors 1, 3, and 7. Matrix $\mathbf{A}_1$ has primaries at row vectors 1, 5, and 7. Matrix $\mathbf{A}_1$ is a linear transform of matrix **A**, but only one primary has been changed. For each matrix, we compute the corresponding **M** and $\mathbf{M}^{-1}$ matrices (Figure 18B). (The reader may choose to compute the corresponding **E** matrices to confirm the invariance of the same primary in both matrix **E** and matrix $\mathbf{E}_1$, in contrast to the high variability of matrix **A** and matrix $\mathbf{A}_1$.)

I am interested in finding the scalar products and angles between the primaries in each matrix. The literature is replete with investigations using matrix **M**, in affine space, for this purpose. However, matrix **M** is highly variable, depending on the choice of all three primaries, and cannot be so employed. On the other hand, the elements or cells of matrix $\mathbf{M}^{-1}$ in Euclidean space are the true scalar products between the primaries, and therefore matrix $\mathbf{M}^{-1}$ must be used. Figure 18 demonstrates invalid calculations of vector lengths and intervening angles using an **M** matrix and also valid calculations using the $\mathbf{M}^{-1}$ matrix—where the vector lengths of the same vector remain constant and the intervening angles between the same vectors remain constant.

I have previously developed relationships between matrix **A** and matrix **E**, and now demonstrate certain parallels with matrix **F**.

There are several basic equations:

$$\mathbf{A}(\mathbf{M}_a)^{-1} = \mathbf{E}, \tag{10.1}$$

$$\mathbf{E}(\mathbf{M}_e)^{-1} = \mathbf{A}, \tag{10.2}$$

$$\mathbf{A}'\mathbf{E} = \mathbf{I}, \tag{10.3}$$

$$\mathbf{E}'\mathbf{A} = \mathbf{I}, \tag{10.4}$$

$$\mathbf{A}\mathbf{E}' = \mathbf{R}, \tag{10.5}$$

$$\mathbf{E}\mathbf{A}' = \mathbf{R}, \tag{10.6}$$

$$\mathbf{F}'\mathbf{F} = \mathbf{I}, \tag{10.7}$$

and

$$\mathbf{F}\mathbf{F}' = \mathbf{R}. \tag{10.8}$$

From any **F** matrix, I can derive an $\mathbf{F}_a$ matrix corresponding to an **A** matrix or an $\mathbf{F}_e$ matrix corresponding to an **E** matrix.

When the primaries correspond to rows in an **F** matrix, the three rows form a 3×3 diagonal matrix where the entries are the respective vector lengths of

A

primary	1.0	0.0	0.0
	3.0	4.0	7.0
primary	0.0	1.0	0.0
	1.0	6.0	9.0
	2.0	2.0	4.0
	7.0	0.0	3.0
primary	0.0	0.0	1.0

$A'A = M$

64.0	22.0	59.0
22.0	57.0	90.0
59.0	90.0	156.0

M^{-1}

0.07610	0.18046	- 0.13289
0.18046	0.62487	- 0.42875
- 0.13289	- 0.42875	0.30403

A_1

primary	1.0	0.0	0.0
	-1.0	2.0	-1.0
	-1.0	0.5	-2.0
	-5.0	3.0	-3.0
primary	0.0	1.0	0.0
	7.0	0.0	3.0
primary	0.0	0.0	1.0

$A_1'A_1 = M_1$

77.0	-17.5	39.0
-17.5	14.25	-12.0
39.0	-12.0	24.0

M_1^{-1}

0.07610	-0.01845	- 0.13289
-0.01845	0.12568	0.09282
-0.13289	-0.09282	0.30403

A

Primaries at 1, 3, 7

Operate on M

8.0000	0.364246	0.590473
0.0000	7.54983	0.954427
0.0000	0.00000	12.49000

	Length		Angle
1	8.000	1,3	71.08°
3	7.549	1,7	53.81°
7	12.49	3,7	17.36°

Operate on M^{-1}

0.27586	0.82751	-0.87365
0.00000	0.79048	-0.98368
0.00000	0.00000	0.55138

Length		Angle
0.2758	1,3	34.15°
0.7904	1,7	150.87°
0.5513	3,7	169.63°

Transform to primaries at 1, 5, 7

Operate on M

8.7749	-0.5283	0.9072
0.0000	3.7749	-0.6488
0.0000	0.0000	4.8989

	Length		Angle
1	8.7749	1,5	121.89°
5	3.7749	1,7	24.88°
7	4.8989	5,7	130.46°

Operate on M^{-1}

0.27586	-0.1886	-0.8736
0.00000	0.3545	0.4748
0.00000	0.0000	0.5513

Length		Angle
0.2758	1,5	100.87°
0.3545	1,7	150.87°
0.5513	5,7	65.65°

Angles NOT preserved

Appropriate Lengths and Angles are preserved

B

Figure 18. An example of the peril associated with computation of scalar products from an **M** matrix in contrast to computations from $\mathbf{M}^{-1}$.

the primaries. To obtain matrix $\mathbf{F}_a$, divide each column in matrix **F** by the corresponding vector length. That operation will place ones in the corresponding rows in matrix **A**—that is, the rows corresponding to the primaries. To obtain matrix $\mathbf{F}_e$, multiply each column in matrix **F** by the corresponding vector length. That operation will place the square of the vector lengths in the corresponding rows in matrix **E**, that is, the rows corresponding to the primaries. Matrix $\mathbf{F}_a$ is a perfectly legitimate **A** matrix where the primaries form a 3×3 identity matrix and matrix $\mathbf{F}_e$ is a perfectly legitimate **E** matrix where the primaries are the entries that appear in the diagonals in matrix **R**.

However, the primaries are innately orthogonal. When we compute matrix $\mathbf{M}_a$ from $\mathbf{F}_a$, that is, $\mathbf{F}'_a\mathbf{F}_a = \mathbf{M}_a$, we obtain a matrix with three entries along the diagonal and zeros elsewhere. Matrix $\mathbf{M}_e = (\mathbf{M}_a)^{-1}$, and we again obtain a matrix with three entries along the diagonal and zeros elsewhere. The entries are the reciprocals of the entries in $\mathbf{M}_a$. Thus, matrices $\mathbf{F}_a$ and $\mathbf{F}_e$ act as the usual **A** and **E** matrices, except that the **M** matrices are diagonal matrices. These are relationships between $\mathbf{F}_a$ and $\mathbf{F}_e$ matrices:

$$(\mathbf{M}_a)^{-1} = \mathbf{M}_e \tag{10.9}$$

and

$$(\mathbf{M}_e)^{-1} = \mathbf{M}_a. \tag{10.10}$$

Now,

$$\mathbf{F}_a(\mathbf{M}_a)^{-1} = \mathbf{F}_e, \tag{10.11}$$

and

$$\mathbf{F}_e(\mathbf{M}_e)^{-1} = \mathbf{F}_a. \tag{10.12}$$

And

$$\mathbf{F}_a(\mathbf{F}'_a\mathbf{F}_a)^{-1}\mathbf{F}'_a = \mathbf{R} \tag{10.13}$$

and

$$\mathbf{F}_e(\mathbf{F}'_e\mathbf{F}_e)^{-1}\mathbf{F}'_e = \mathbf{R} \tag{10.14}$$

$$\mathbf{F}_e\mathbf{F}'_a = \mathbf{R}. \tag{10.15}$$

Similarly,

$$\mathbf{F}_a\mathbf{F}'_e = \mathbf{R}. \tag{10.16}$$

Then,

$$\mathbf{F}'_a\mathbf{F}_e = \mathbf{I} \tag{10.17}$$

and

$$\mathbf{F}'_e\mathbf{F}_a = \mathbf{I}. \tag{10.18}$$

Given an **F** orthonormal matrix, we may derive the corresponding orthogonal (but not normalized) $\mathbf{F}_a$ matrices by dividing each column by its vector length, and we may derive the corresponding orthogonal (but not normalized) $\mathbf{F}_e$ matrix by multiplying each column by its vector length. In the **F** matrix, the vector lengths are generally different for each column. This means that the **F**, $\mathbf{F}_a$, and $\mathbf{F}_e$ matrices have precisely the same basic configuration but are scaled differently. Figure 19 illustrates three orthonormal functions in an **F** matrix, referring to its own scaling on the y-axis. The same three orthonormal functions also refer to the $\mathbf{F}_a$ and $\mathbf{F}_e$ scaling in their y-axes; the $\mathbf{F}_a$ and $\mathbf{F}_e$ matrices remain orthogonal when referred to their respective scales, but they are no longer normalized. This relationship between **A**, **E**, and **F** matrices will be demonstrated later (see Figure 24).

When we are given an **F** matrix where the primaries do not correspond to rows and the vector lengths of the row primaries are unknown, then the **F** matrix may be configured into an infinite number of **A** matrices and their corresponding **E** matrices, where equations 10.1 through 10.8 hold. The normalization procedures, of Gram-Schmidt and those inherent in matrix **R** operations, result in loss of information concerning the original **A** and **E** matrices; the problem is to find sufficient constraints, as is often the case in the sciences, but an external criterion may be evoked to isolate single, special **A** and **E** matrices.

Although **A**, **E**, and **F** matrices are related by the equations developed, the matrices have very different properties. This has not been understood in the literature. For example, MacAdam (1954) uses the term *color-matching functions* in reference to the columns in both matrix **A** and matrix **F**. The columns in matrix **A** are indeed color-matching functions and would emerge in a color-matching experiment; the rows corresponding to primaries comprise an identity matrix. The columns of matrix **F** are *not* color-matching functions and cannot emerge from a color-matching experiment; the row primaries comprise a diagonal matrix, not an identity matrix. The columns of matrix **F** are termed *tricolor functions.*

Indeed, the **M** matrices of matrix **A**, matrix **E**, and matrix **F** are very different. Using matrix **A**, the **M** matrix may have any entries. Using matrix **E**, the **M** matrix comprises scalar products that are subsections of matrix **R**. Using matrix **F**, the **M** matrix is always an identity matrix.

11

Linear Transformation of Color Stimuli

The linear transformation of colorimetric data, from one triplet of color-matching functions to another, has long been a festering problem. With the introduction of the construct of the fundamental, a similar problem exists for the transformation of one triplet of transformations to another. The same is true for orthonormal functions. Matrix-**R** operations provide direct solutions without assumptions beyond those inherent in Grassmann's laws.

I consider tranformations between matrices **A**, **E**, and **F**, or between any combination. These transformations are illustrated in Figure 19.

Transformation I: Color-Matching Functions (A Matrices)

The triplet of color-matching functions, emerging from a color-matching experiment, are unique to the triplet of primaries used in the experiment. This triplet of color-matching functions comprises the columns of an **A** matrix, while the tristimulus values of each monochromatic standard are one row in an **A** matrix. We may, of course, compute the tristimulus values **Q** of any radiometric function **N** by the usual equation, $\mathbf{Q} = \mathbf{A}'\mathbf{N}$.

For almost a century, colorimetry has been occupied with the solution of an axiomatic problem. Given a triplet of color-matching functions from one triplet of primaries, what are the color-matching functions that would emerge using three different primaries? Herbert E. Ives (1915, 1923) appears to have first

Matrix A_1

0.1219	0.0340	0.2194
0.1052	-0.0428	0.1390
0.2045	-0.3668	-0.1085
-0.4496	0.1018	0.2928
-0.1611	-0.0946	-0.0523
-0.5545	0.3052	0.2298
-0.5429	0.4933	0.2885

$T_{12} = E'_1 A_2$

-0.1176	-0.0588	0.4118
-0.0588	0.4706	-0.2941
0.4706	-0.7647	0.3529

$T_{21} = E'_2 A_1$

1.0000	5.0000	3.0000
2.0000	4.0000	1.0000
3.0000	2.0000	1.0000

Matrix A_2

0.0869	-0.1589	0.1176
0.0556	-0.1326	0.1050
-0.0535	-0.1017	0.1538
0.1847	-0.1496	-0.1117
-0.0001	0.0050	-0.0570
0.1554	0.0005	-0.2370
0.1706	0.0435	-0.2668

Matrix E_1

1.0000	0.0000	2.0000
0.5455	-0.3636	1.5000
-0.5455	-1.6364	0.5000
-0.8182	-1.4545	1.3000
-1.0000	-1.0000	-0.5000
-0.7273	-0.1818	0.0000
0.0000	1.0000	0.0000

$T_{12} = A'_1 E_2$

1.0000	2.0000	3.0000
5.0000	4.0000	2.0000
3.0000	1.0000	1.0000

$T_{21} = A'_2 E_1$

-0.1176	-0.0588	0.4706
-0.0588	0.4706	-0.7647
0.4118	-0.2941	0.3529

Matrix E_2

7.0000	4.0000	5.0000
3.2275	1.1366	2.4093
-7.2275	-7.1366	-4.4093
-4.1907	-6.1544	-4.0636
-7.5000	-6.5000	-5.5000
-1.6363	-2.1818	-2.5455
5.0000	4.0000	2.0000

Matrix F_1

0.5134	-0.3724	0.3981
0.2801	-0.3734	0.2523
-0.2801	-0.5632	-0.1968
-0.4201	-0.3765	0.5313
-0.5134	-0.0959	-0.0948
-0.3734	0.1857	0.4170
-0.0000	0.4683	0.5234

$T_{12} = F'_1 F_2$

0.5418	0.7641	0.3502
-0.3930	0.5986	-0.6980
-0.7430	0.2406	0.6246

$T_{21} = F'_2 F_1$

0.5418	-0.3930	-0.7430
0.7641	0.5986	0.2405
0.3502	-0.6980	0.6246

Matrix F_2

0.1287	0.2652	0.6884
0.1110	0.0512	0.5163
0.2158	-0.5985	0.1721
-0.4744	-0.4185	0.4475
-0.1700	-0.4725	-0.1721
-0.5851	-0.0739	0.0000
-0.5729	0.4062	0.0000

Figure 19. Transformation of colorimetric data.

addressed the issue in 1915, using both algebra and geometry. For Ives, the problem was phrased with respect to two colorimeters. Given the specification of a color stimulus by a match with the primaries of one colorimeter, what is the specification of the same color stimulus by a match with different primaries in a different colorimeter? Ives employed several marginal assumptions—that a white sensation was evoked by equal quantities of each primary in the first triplet, and similarly for the second triplet. Furthermore, Ives assumed that the solution depended on the relative luminosities of the two primary triplets. Numerous colorimetrists have attacked the problem with variations, including Froelich (1924), Guild (1924–1925a, 1924–1925b), Dziobek (1926), Runge (1928), Judd (1930), Smith (1931–1932), Smith and Guild (1931–1932), MacAdam (1937, 1942, 1953, 1954), Wintringham (1951), Bingley (1954), Howett and Graham (1957), Graham (1965), and Pratt (1978).

I now present an exact solution derived explicitly from matrix-**R** operations and requiring no assumptions whatever (see Figure 19, top row). The problem may be stated formally. Given a $k \times 3$ matrix $\mathbf{A}_1$ (three color-matching functions, the first triplet of primaries) and a linear transform, $k \times 3$ matrix $\mathbf{A}_2$ (three different color-matching functions, the second triplet of primaries). The first set of primaries has tristimulus values in the frame of the second (3×3 matrix $\mathbf{Q}_{12}$), and the second set of primaries has tristimulus values in the frame of the first (3×3 matrix $\mathbf{Q}_{21}$). Note that the columns of matrix $\mathbf{A}_1$ are color-matching functions and in general are not the radiometric functions or fundamentals of the primaries used in the color-matching experiment. Designate the transformation matrices as 3×3 nonsingular matrix $\mathbf{T}_{12}$ carrying matrix $\mathbf{A}_1$ into matrix $\mathbf{A}_2$, and 3×3 nonsingular matrix $\mathbf{T}_{21}$ carrying matrix $\mathbf{A}_2$ into matrix $\mathbf{A}_1$. As will be demonstrated momentarily, $\mathbf{T}_{12} = \mathbf{Q}_{12}$ and $\mathbf{T}_{21} = \mathbf{Q}_{21}$. The problem: Given $\mathbf{A}_1$ and $\mathbf{A}_2$ alone, derive matrix $\mathbf{Q}_{12}$ and matrix $\mathbf{Q}_{21}$.

Thus stated, the solution becomes immediately manifest. We have two triplets of color-matching functions, $\mathbf{A}_1$ and $\mathbf{A}_2$. In the first triplet, each primary has an immutable fundamental (one column in $k \times 3$ matrix $\mathbf{E}_1$). In the second triplet, each primary also has an immutable fundamental (one column in $k \times 3$ matrix $\mathbf{E}_2$). By equations 7.6 and 7.13,

$$\mathbf{A}_1(\mathbf{A}_1'\mathbf{A}_1)^{-1} = \mathbf{A}_1\mathbf{M}_{1e} = \mathbf{E}_1, \tag{11.1}$$

and

$$\mathbf{A}_2(\mathbf{A}_2'\mathbf{A}_2)^{-1} = \mathbf{A}_2\mathbf{M}_{2e} = \mathbf{E}_2. \tag{11.2}$$

We seek matrix $\mathbf{Q}_{12}$, the tristimulus values of the primaries in $\mathbf{A}_1$ with reference to the primaries in $\mathbf{A}_2$. Better still, we seek the tristimulus values of the fundamentals (with or without added residuals) of the primaries used in the determination of the color-matching functions in $\mathbf{A}_1$ (represented in matrix $\mathbf{E}_1$) with reference to the color-matching functions of the primaries in $\mathbf{A}_2$. Similarly for matrix $\mathbf{Q}_{21}$. By equation 7.1, the usual definition of tristimulus values, and by equations 7.17 and 7.18,

$$\mathbf{E}_1'\mathbf{A}_2 = \mathbf{A}_2'\mathbf{E}_1 = \mathbf{Q}_{12} = \mathbf{T}_{12} \tag{11.3}$$

and

$$\mathbf{E}_2'\mathbf{A}_1 = \mathbf{A}_1'\mathbf{E}_2 = \mathbf{Q}_{21} = \mathbf{T}_{21}. \tag{11.4}$$

These equations, invoking matrix-**R** operations, are powerful, and solve the problem directly.

Several years ago, following a discussion of these transformations at the Matrix **R** Conference (Fairman, 1986), Hugh S. Fairman (personal communication, 1987) declared that the problem may be solved without overt discussion of matrix-**R** decomposition.

By the definition of colorimetric transformation,

$$\mathbf{A}_1\mathbf{T}_{12} = \mathbf{A}_2. \tag{11.5}$$

Define,

$$\mathbf{E}_1 = \mathbf{A}_1(\mathbf{A}_1'\mathbf{A}_1)^{-1} \tag{11.6}$$

as per equation 7.12. By a theorem in linear algebra,

$$\mathbf{E}_1' = (\mathbf{A}_1'\mathbf{A}_1)^{-1}\mathbf{A}_1'. \tag{11.7}$$

Premultiply both sides of equation 11.5 by $\mathbf{E}_1'$,

$$\mathbf{E}_1'\mathbf{A}_1\mathbf{T}_{12} = \mathbf{E}_1'\mathbf{A}_2. \tag{11.8}$$

Substituting equation 11.7 into equation 11.8,

$$(\mathbf{A}_1'\mathbf{A}_1)^{-1}\mathbf{A}_1'\mathbf{A}_1\mathbf{T}_{12} = \mathbf{E}_1'\mathbf{A}_2. \tag{11.9}$$

Thus,

$$\mathbf{IT} = \mathbf{T} = \mathbf{E}_1'\mathbf{A}_2, \tag{11.10}$$

and we have

$$\mathbf{A}_1(\mathbf{E}_1'\mathbf{A}_2) = \mathbf{A}_2. \tag{11.11}$$

Fairman (personal communication, 1986) wrote: "So the colorist, given $\mathbf{A}_1$ and $\mathbf{A}_2$, calculates $\mathbf{E}_1$, . . . multiplies its transpose into $\mathbf{A}_2$ and gets [the transformation matrix **T**]. [The colorist needs know nothing of matrix **R** operations], but has solved a hitherto impossible problem." The transformation matrix may be written exclusively with **A** matrices,

$$\mathbf{A}_1[(\mathbf{A}_1'\mathbf{A}_1)^{-1}(\mathbf{A}_1'\mathbf{A}_2)] = \mathbf{A}_2, \tag{11.12}$$

where $[(\mathbf{A}_1'\mathbf{A}_1)^{-1}(\mathbf{A}_1'\mathbf{A}_2)] = \mathbf{T}_{12}$. Similar proofs produce matrix $\mathbf{T}_{21}$, the transformation from $\mathbf{A}_2$ to $\mathbf{A}_1$.

These equations permit the answer to another question: What are the tristimulus values ($\mathbf{Q}_{11}$ and $\mathbf{Q}_{22}$) of the fundamental metamers of a triplet of primaries with reference to the color-matching functions of the same primaries?

By equations 10.3 and 10.4, the solution is

$$\mathbf{E}_1'\mathbf{A}_1 = \mathbf{A}_1'\mathbf{E}_1 = \mathbf{I}, \qquad (11.13)$$

and

$$\mathbf{E}_2'\mathbf{A}_2 = \mathbf{A}_2'\mathbf{E}_2 = \mathbf{I}. \qquad (11.14)$$

Matrix **I** is an identity matrix, and the tristimulus values are (1,0,0), (0,1,0), and (0,0,1). Because the tristimulus values are merely dimensionless weights on the fundamentals of the primaries, the fundamental of each primary is weighted by unity (the other two fundamental metamers weighted by zero) to reproduce the fundamental. This identity matrix, in fact, is the three rows in an **A** matrix that correspond to the primaries.

Long ago, a simple variation of this transformation matrix, with no assumptions, was given by Cohen and Walker (1946). This solution is viable only when the primaries of matrix $\mathbf{A}_2$ appear as rows in matrix $\mathbf{A}_1$. Now, the primaries of matrix $\mathbf{A}_1$ appear as an identity matrix in matrix $\mathbf{A}_1$. Three different rows in matrix $\mathbf{A}_1$, however, correspond to the primaries in matrix $\mathbf{A}_2$; they are the tristimulus values of the primaries in matrix $\mathbf{A}_2$ with respect to the primaries in $\mathbf{A}_1$. The corresponding rows in matrix $\mathbf{A}_2$, however, are primaries in matrix $\mathbf{A}_2$ and therefore comprise an identity matrix in matrix $\mathbf{A}_2$. Therefore, transformation $\mathbf{T}_{12}$ is the inverse of these rows in matrix $\mathbf{A}_1$. The same is true for the transformation matrix $\mathbf{T}_{21}$.

I emphasize that **A** matrices are affine (the vectors are in affine space) and are transformed by an affine transformation. Given $\mathbf{A}_1$ and $\mathbf{A}_2$, where two primaries are identical in the triplets of both matrices, in general, all three columns of matrix $\mathbf{A}_2$ will differ from the corresponding columns of matrix $\mathbf{A}_1$, although two primaries are identical in both matrices. This has already been discussed and illustrated.

Transformation II: Fundamental Functions (E Matrices)

The three columns of an **E** matrix are the fundamentals of the primaries used in a color-matching experiment (see Figure 19, second row). The color-matching functions emerge from the experiment. Given an **A** matrix, the corresponding matrix **E** may be computed.

The problem is to find a 3×3 nonsingular matrix $\mathbf{T}_{12}$ that transforms $k \times 3$

matrix $\mathbf{E}_1$ with one triplet of primary fundamentals to another $k \times 3$ matrix $\mathbf{E}_2$ with a second triplet of primary fundamentals. The entries of **E** matrices are scalar products.

Given matrices $\mathbf{E}_1$ and $\mathbf{E}_2$, we seek matrix $\mathbf{T}_{12}$, the transformation carrying matrix $\mathbf{E}_1$ to $\mathbf{E}_2$. Then,

$$\mathbf{E}_1\mathbf{T}_{12} = \mathbf{E}_2. \tag{11.15}$$

From equation 7.12,

$$\mathbf{E}_1 = \mathbf{A}_1(\mathbf{A}_1'\mathbf{A}_1)^{-1}, \tag{11.16}$$

and

$$\mathbf{E}_2 = \mathbf{A}_2(\mathbf{A}_2'\mathbf{A}_2)^{-1}. \tag{11.17}$$

Therefore,

$$\mathbf{A}_1(\mathbf{A}_1'\mathbf{A}_1)^{-1}\mathbf{T}_{12} = \mathbf{A}_2(\mathbf{A}_2'\mathbf{A}_2)^{-1}. \tag{11.18}$$

Multiply both sides of the equation by $\mathbf{A}_1'$:

$$\mathbf{A}_1'\mathbf{A}_1(\mathbf{A}_1'\mathbf{A}_1)^{-1}\mathbf{T}_{12} = \mathbf{A}_1'\mathbf{A}_2(\mathbf{A}_2'\mathbf{A}_2)^{-1}. \tag{11.19}$$

Thus,

$$\mathbf{I}\mathbf{T}_{12} = \mathbf{T}_{12} = \mathbf{A}_1'\mathbf{E}_2. \tag{11.20}$$

Therefore,

$$\mathbf{E}_1(\mathbf{A}_1'\mathbf{E}_2) = \mathbf{E}_2. \tag{11.21}$$

The transformation matrix may be written entirely in **E** matrices:

$$\mathbf{E}_1[(\mathbf{E}_1'\mathbf{E}_1)^{-1}(\mathbf{E}_1'\mathbf{E}_2)] = \mathbf{E}_2. \tag{11.22}$$

Similar proofs produce matrix $\mathbf{T}_{21}$, the transformation from $\mathbf{E}_2$ to $\mathbf{E}_1$. Note the symmetry of transformations for **A** matrices and the transformation for **E** matrices. For **A** matrices, $\mathbf{T}_{12} = \mathbf{E}_1'\mathbf{A}_2$, and for **E** matrices, $\mathbf{T}_{12} = \mathbf{A}_1'\mathbf{E}_2$.

Although the equations for the transformation of **A** matrices are similar to the equations for the transformation of **E** matrices, the results are very different. Suppose that we have an $\mathbf{E}_1$ matrix with the fundamentals of three primaries used in the color-matching experiment. We transform to another matrix $\mathbf{E}_2$, where two primaries are identical to those in matrix $\mathbf{E}_1$. The fundamentals for the two common primaries will always be identical; the column for the third primary will be different. The invariance of the fundamentals contrasts sharply with the highly variable character of the color-matching primaries in an **A** matrix, although the primaries are identical.

I have already demonstrated that $\mathbf{E}_1(\mathbf{M}_{1e})^{-1}\mathbf{E}_1' = \mathbf{R}$. We can also demonstrate that $\mathbf{E}_1(\mathbf{M}_{12})^{-1}\mathbf{E}_2 = \mathbf{R}$, where $\mathbf{M}_{12}$ is an **M** matrix, the scalar products between the fundamentals in the two sets of **E** matrices. For this calculation, $\mathbf{M} = (\mathbf{E}_1'\mathbf{E}_2)'$, or to emphasize the transpose, $\mathbf{M}' = (\mathbf{E}_1'\mathbf{E}_2)$.

Transformation III: Tricolor Orthonormal Functions (F Matrices)

Orthonormal functions or **F** matrices have been discussed previously. The columns represent the orthogonal and normalized reference primaries in fundamental color space (see Figure 19, third row). The rows represent the projections of any vector, signifying one fundamental and its evoked color sensation, onto the three reference primaries. Here, the rules of Euclidean vector transformations apply absolutely.

One **F** matrix is transformed to another by so-called orthogonal transformation **Z** matrices, not to be confused with the orthogonal **F** matrices themselves. The determinant of any **Z** matrix is either $+1$ or -1, and the entries are the direction cosines (or simply cosines) between the reference frame in $\mathbf{F}_1$ and the reference frame in $\mathbf{F}_2$. (Statisticians will recognize the direction cosines as the geometric interpretation of the Pearson product-moment correlation coefficients.)

In two dimensions, the transformation matrix is of the form

$$\mathbf{Z}_{12} = \begin{bmatrix} \cos(\phi) & -\sin(\phi) \\ \sin(\phi) & \cos(\phi) \end{bmatrix}, \tag{11.23}$$

where ϕ is the angle between the axes in $\mathbf{F}_1$ and the corresponding axes in $\mathbf{F}_2$. Although cos and sin functions appear to be intermixed, all entries are actually cosines (direction cosines) depending on the measurement of the intervening angle; for example, $\cos 45° = 0.7071$ and $\cos(180° + 45°) = -0.7071$.

In three-dimensional orthonormal **F** matrices, the problem is essentially equivalent, and the nine entries in the transformation matrix are the direction cosines between the axes in the first orthogonal triplet and the corresponding axes in the second orthogonal triplet. Again the transformation matrix is an orthogonal matrix, that is, its transpose equals its inverse and its determinant is either $+1$ or -1. This is well demonstrated and explained by Snyder and Sisam (1941) with direction cosines and Ayre, Stephens, and Mock (1967) with scalar products. However, a succession of two-dimensional transformations is the equivalent of one three-dimensional transformation.

Given matrices $\mathbf{F}_1$ and $\mathbf{F}_2$, we may find the transformation matrix $\mathbf{Z}_{12}$ by matrix-**R** operations. Matrices $\mathbf{FF}' = \mathbf{R}$, and $\mathbf{F}'\mathbf{F} = \mathbf{I}$. This last equation states that the direction cosines between the reference axes are either zero, $\cos\phi = 1$, or one, $\cos\phi = 0$. We may multiply $\mathbf{F}_1'\mathbf{F}_2$ to obtain the direction cosines between the orthogonal frame in $\mathbf{F}_1$ and the orthogonal frame in $\mathbf{F}_2$. The 3×3 matrix $\mathbf{Z}_{12}$ is the transformation matrix between $\mathbf{F}_1$ and $\mathbf{F}_2$. This is illustrated in Figure 19.

We may write the transformation matrix from $\mathbf{F}_1$ to $\mathbf{F}_2$ entirely in $\mathbf{F}$ matrices:

$$\mathbf{F}_1[\mathbf{F}_1'\mathbf{F}_2] = \mathbf{F}_2, \tag{11.24}$$

where $\mathbf{Z}_{12} = \mathbf{F}_1'\mathbf{F}_2$.

In transforming from one orthogonal reference frame to another, the vector lengths and angles remain fixed. We are merely referring these same vectors to a new and different frame; this contrasts with the transformation of, say, x_t, y_t, z_t vectors in affine space, where movement of the axes brings a collateral movement of all vectors with respect to themselves.

Thus far, we have attended to the transformation of one orthogonal reference triplet to another orthogonal reference triplet, that is, $\mathbf{F}_1$ to $\mathbf{F}_2$. Fundamental color space, however, suggests that the reference frame need not be orthogonal with the angles between reference axes at 90 degrees; as will be demonstrated later, almost all color theories refer not to an orthogonal frame but rather to an oblique frame where the reference angles are not at 90 degrees.

Philosophers of science have long debated whether the fundamental constructs of any discipline ought to be independent, because constructs are merely arbitrary devices for understanding phenomena observed in nature. When the constructs are independent, then the errors of one variable are not inherently attributable to (or correlated with) errors of another variable; for example, the NTSC primaries are highly correlated with each other, and this greatly contributes to the apparent "color shimmering" of uniformly colored fields observed on color television receivers. By our reasoning, orthonormal primaries are to be preferred; furthermore, their mathematics is simplified.

Nevertheless, we address the theoretical mathematical problem of transforming from an orthonormal frame to an oblique frame. Mathematicians do not have a designation for a matrix with oblique, normalized reference axes, a name that would parallel *orthonormal*—perhaps *oblinormal* (although that may be a contradiction in terms). Transformations to nonnormalized oblique frames, however, have long been studied by mathematicians using so-called Euler angles, after Leonhard Euler, a Swiss mathematician of the eighteenth century. (See Boyer, 1956, and Bailey & Woods, 1898.)

Transformation IV: Generalized Functions

All transformations above may be represented by a single, generalized equation. Here is a basic matrix-$\mathbf{R}$ equation:

$$\mathbf{A}(\mathbf{A}'\mathbf{A})^{-1}\mathbf{A}'\mathbf{N} = \mathbf{N}^*. \tag{11.25}$$

This equation may be generalized to give the **T** (or **Z**) matrix transforming any **A** matrix to any other **A** matrix, any **E** matrix to any other **E** matrix, and any **F** matrix to any other **F** matrix. Furthermore, it will give the **T** matrix transforming any **A** matrix to any **E** or **F** matrix, any **E** matrix to any **A** or **F** matrix, and any **F** matrix to any **A** or **E** matrix. The transformation is independent of the choice of primaries, which may be very different in the several diverse **A**, **E**, and **F** matrices.

This is the equation in its generalized transformation form, (Cohen & Kappauf, 1985):

$$\mathbf{D}[(\mathbf{D}'\mathbf{D})^{-1}\mathbf{D}'\mathbf{C}] = \mathbf{C}, \tag{11.26}$$

where $\mathbf{T} = [(\mathbf{D}'\mathbf{D})^{-1}\mathbf{D}'\mathbf{C}]$. For example, let $\mathbf{D} = \mathbf{F}$ and $\mathbf{C} = \mathbf{A}$. Then the transformation between matrix **F** and matrix **A** is given by $\mathbf{T} = [(\mathbf{F}'\mathbf{F})^{-1}\mathbf{F}'\mathbf{A}]$.

12

Two Preferred Reference Frames in Fundamental Color Space

The vector configuration of fundamental color space is predetermined and cannot be altered by an investigator. The selection of reference axes (primaries), however, is discretionary. I have an inherent preference for orthogonal frames, in contrast to oblique frames, because the mathematics is less cumbersome. If *Natura est simplex*, as Newton insisted, then the frame should be orthogonal. Actually, all classical theories of color have oblique frames, although this was not realized by their creators. Furthermore, the CIE R,G,B and X,Y,Z reference frames are oblique—indeed, very oblique.

Here, I discuss two important reference frames and their internal characteristics. Two additional reference frames are more properly addressed with respect to complementary color stimuli and are discussed later.

The Canonical Configuration

The canonical configuration of the manifest reference frame is illustrated in Figure 14B, where the first axis is coincident with the equal-energy vector at the centroid of the configuration, and the remaining two axes are on a plane orthogonal to the first axis and also orthogonal to each other The first axis may properly be denominated the *achromatic axis*; this reference axis represents a unique "white" sensation whose components are the collected standards of the color-matching experiment. The orthogonal plane may properly be denom-

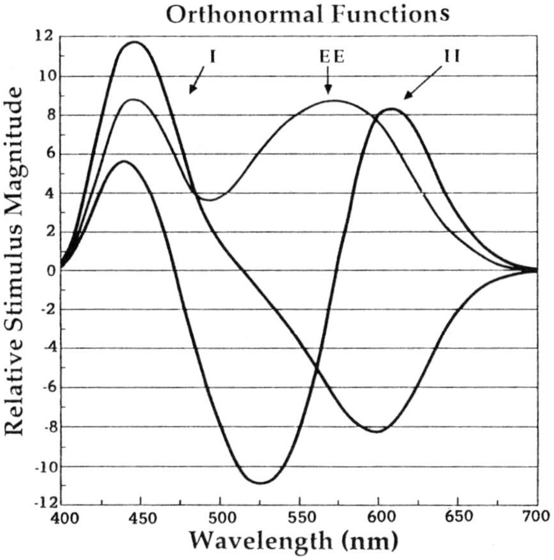

Figure 20. The three canonical orthonormal functions: Columns of an intriguing **F** matrix.

inated the *chromatic plane*; the two reference axes on this plane have a zero achromatic component. The position of the two chromatic vectors on the chromatic plane is arbitrary, but several positions may have singular advantages. Figure 20 depicts a typical triplet of canonical orthonormal tricolor functions. The first axis has been placed through the equal-energy vector at the centroid of the fundamental vector configuration; this may be termed the *achromatic axis* because it represents only achromatic color stimuli and achromatic sensations. The function exhibits the now-familiar two-modes of equal magnitude. Axes I and II are orthogonal to each other and lie on the chromatic plane, orthogonal plane to the chromatic axis. (Axes I and II may be rotated orthogonally on that plane.) Axes I and II have a remarkable property: the algebraic sum of the areas under each function is precisely zero. This may be affirmed by calculation or by visual inspection of the functions.

This canonical orthonormal configuration has an astonishing property that may be of critical importance in the transmission of color information. The canonical orthonormal tricolor functions comprise an **F** matrix, and as we expect, $\mathbf{F'F} = \mathbf{I}$, meaning that the sums of squares of the columns equal unity and the sum of "cross-products" between columns equals zero, again as we expect. The next property, however, is totally unexpected. Not only are the sums of squares of each column equal to unity, but the algebraic sums of the second and third columns (those orthogonal to each other and on the chromatic plane) are both precisely zero. This is true in every instance where the first axis passes through the equal-energy vector, irrespective of the orthogonal rotational position of the second and third axes on the chromatic plane. Thus, in Figure 14B, the algebraic sum of axis I is precisely zero, and the same is true for axis II. This property is not characteristic of other orthonormal tricolor functions as, for example, a triplet with the first column proportional to the luminance function (See MacAdam, 1954, and Burns, Cohen, & Kuznetsov, 1990.)

The reason for this "zero addition" may be found in the theorem, previously discussed, that sums of columns in matrix **A** are the tristimulus values of an equation that matches the color sensation evoked by an equal-energy source. Recall that every **F** matrix may theoretically be resolved into an $\mathbf{F}_a$ and $\mathbf{F}_e$ matrix, where the columns of $\mathbf{F}_a$ and **F** are proportional and the columns of $\mathbf{F}_e$ and **F** are proportional. With respect to a set of orthogonal functions, where the first column is proportional to the fundamental for an equal-energy stimulus, the algebraic sums of the columns are (J, 0, 0), where J is a factor that will modify the fundamental in the first column and the zeros modify the second and third columns. Thus, in the color-matching equation, the fundamental of the first column (the fundamental of equal-energy stimulus) is weighted by factor J, and the remaining two columns are each weighted by zero, so that the sum of the weighted fundamentals is the weighted fundamental of an equal-energy stimulus.

Within this canonical fundamental color space, each color stimulus and sensation is represented by a unique vector, and additive color mixture is the addition of these vectors. Each vector is specified by three coordinates, so that the coordinates of the resultant vector are merely the sums of the respective coordinates, as is always the case in orthogonal frames.

A given fundamental color vector therefore has its first projection onto the "white" achromatic axis and the second two projections onto the chromatic axes lying on the chromatic plane. Because the total norm (length) of the vector is the root of the sums of squares of the projections, we may resolve any fundamental color vector into two components: the projection onto the first achromatic axis (the achromatic norm) and the vector sum of the chromatic components in the chromatic plane (the chromatic norm). The final chromatic norm is independent of the chromatic axes position on the chromatic plane.

An interesting strategy develops using this particular **F** matrix to compute matrix **R**. We specify the columns as $\mathbf{F}_1$, $\mathbf{F}_2$, and $\mathbf{F}_3$. Column $\mathbf{F}_1$ is the equal-energy achromatic fundamental, and columns $\mathbf{F}_2$ and $\mathbf{F}_3$ are arbitrary orthogonal chromatic fundamentals. We compute matrix **R** progressively:

$$\mathbf{FF'} = \mathbf{R}, \tag{12.1}$$

$$\mathbf{F}_1\mathbf{F}_1' = \mathbf{R}_a, \tag{12.2}$$

$$[\mathbf{F}_2\mathbf{F}_2' + \mathbf{F}_3\mathbf{F}_3'] = \mathbf{R}_c, \tag{12.3}$$

$$\mathbf{R}_a + \mathbf{R}_c = \mathbf{F}_1\mathbf{F}_1' + [\mathbf{F}_2\mathbf{F}_2' + \mathbf{F}_3\mathbf{F}_3'] = \mathbf{R}. \tag{12.4}$$

We have split matrix **R** into two additive parts. Figure 21A is a computer rendering of the achromatic portion of matrix **R**. Figure 21B is a computer rendering of the the chromatic portion of matrix **R**. The illustrations show the achromatic and chromatic components of spectral, monochromatic stimuli; the components have many negative values but are theoretically important.

The achromatic component and the chromatic components sum to matrix **R**, as shown in Figure 5. The achromatic component is the first column of the canonical orthonormal matrix **F** and represents the familiar two-peaked equal-energy fundamental, displayed in three dimensions in Figure 21A. The achromatic component is the second two columns of the canonical matrix **F** (representing those two columns whose algebraic sum is always zero), here displayed in three dimensions. Thus, the columns and rows of the chromatic component in *B* invariably add algebraically to zero. The product of the matrix in *A* and the matrix in *B* is a $k \times k$ null matrix, because within the same monochromatic color stimulus, the achromatic component is always orthogonal to the chromatic component. These component matrices may be used to isolate the achromatic and chromatic components of heterochromatic fundamentals. See Figure 22.

Because of the orthogonal structure of matrices $\mathbf{R}_a$ and $\mathbf{R}_c$ and the property that the columns in $\mathbf{F}_2$ and $\mathbf{F}_3$ always sum algebraically to zero, another interesting characteristic of matrix **R** emerges:

$$\mathbf{R}_a\mathbf{R}_c = \mathbf{0}. \tag{12.5}$$

Given a radiometric function **N**, we may divide the fundamental $\mathbf{N}^*$ into scalar achromatic component $\mathbf{N}_a$ and scalar chromatic components $\mathbf{N}_c$, where

$$\mathbf{N}_a + \mathbf{N}_c = \mathbf{N}^*. \tag{12.6}$$

Thus,

$$\mathbf{R}_a\mathbf{N} = \mathbf{R}_a\mathbf{N}^* = \mathbf{N}_a^* \tag{12.7}$$

and

$$\mathbf{R}_c\mathbf{N} = \mathbf{R}_c\mathbf{N}^* = \mathbf{N}_c^*. \tag{12.8}$$

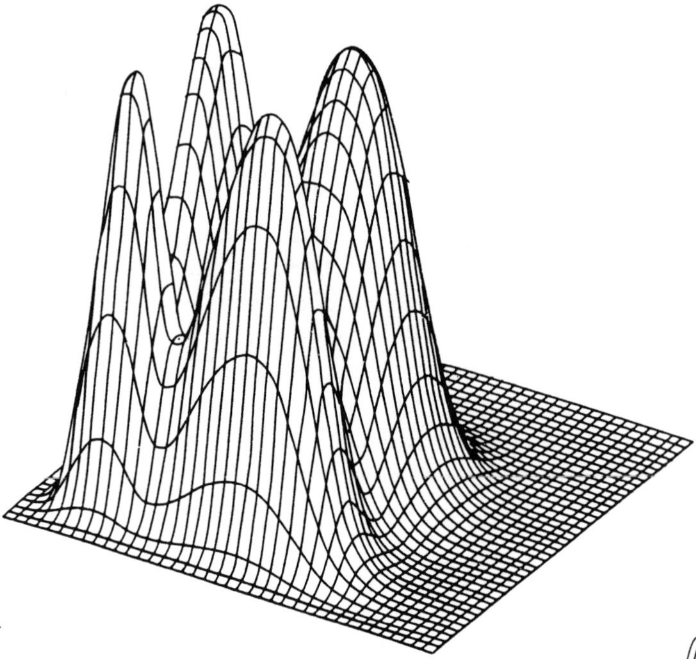

A

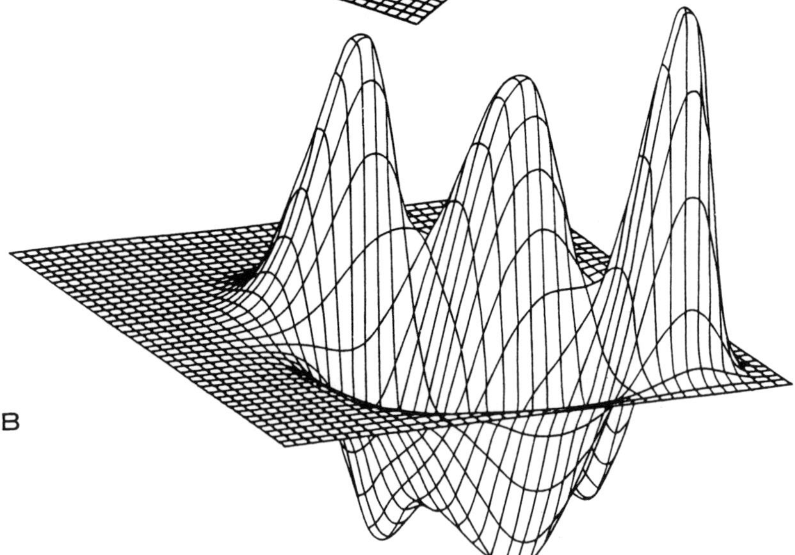

B

Figure 21. Computer renderings of achromatic and chromatic components of matrix **R**.

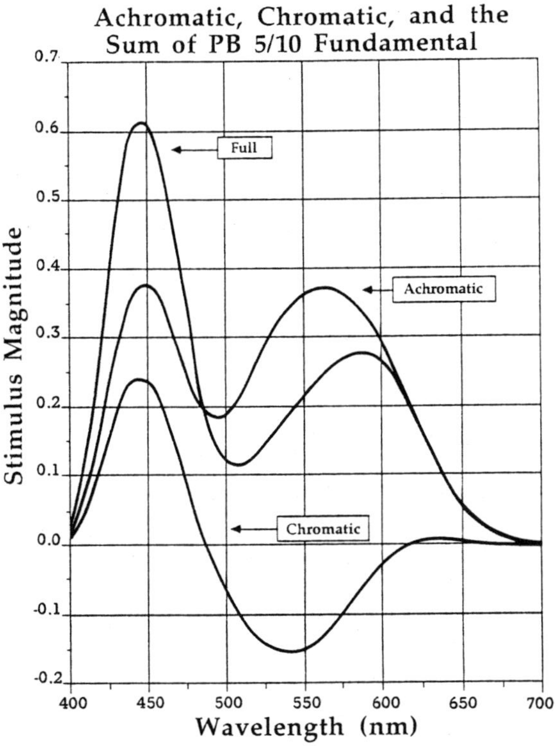

Figure 22. The fundamental of Munsell PB 5/10 resolved into its achromatic and chromatic components.

Figure 22 illustrates the fundamental of a Munsell chip divided into its achromatic and chromatic components. The resolution has been accomplished by separately premultiplying the Munsell fundamental by the achromatic and chromatic components of matrix **R** shown in Figure 21. The achromatic and chromatic components of any fundamental are always uncorrelated.

By the theorem developed above, the algebraic sum of $\mathbf{N}_c^*$ must always equal zero. Again, this may be of considerable import in the transmission of colorimetric information.

I address the question: Why select the equal-energy source as the "white" for the achromatic component? After all, equal energy is not a popular illuminant, and not necessarily preferred, say, as a substitute for daylight. In fact, we can select any other illuminant; the direct transformation matrices from the 1964 X, Y, Z data are given for several standard illuminants in Figure 23. Matrix $\mathbf{T}_E$, for example, transforms the CIE data to an orthonormal **F** ma-

Equal-energy illuminant:

$$\begin{bmatrix} .09947 & -.15392 & -.55609 \\ .20105 & .33414 & .41574 \\ .11061 & -.18021 & .14036 \end{bmatrix} = T_E$$

Standard illuminant A:

$$\begin{bmatrix} .26223 & -.50310 & -.14136 \\ .09997 & .55705 & .06785 \\ -.00326 & .00626 & .25370 \end{bmatrix} = T_A$$

Standard Illuminant B:

$$\begin{bmatrix} .10754 & -.20822 & -.53566 \\ .20677 & .35394 & .39608 \\ .09024 & -.17471 & .16046 \end{bmatrix} = T_B$$

Standard illuminant C:

$$\begin{bmatrix} .05353 & -.07576 & -.57728 \\ .21722 & .29328 & .43845 \\ .13305 & -.18828 & .10611 \end{bmatrix} = T_C$$

Standard illuminant D_{65}:

$$\begin{bmatrix} .04311 & -.06715 & -.57921 \\ .23534 & .27413 & .44088 \\ .12580 & -.19597 & .10091 \end{bmatrix} = T_D[65]$$

Figure 23. Direct transformation matrices from the 1964 CIE X,Y,Z color-matching functions to orthonormal **F** matrices.

trix where the first column is proportional to the equal-energy fundamental and the remaining two axes are on a plane orthogonal to the first axis; this is the canonical orthonormal matrix of this book. The remaining matrices transform to orthonormal **F** matrices where the first column is proportional respectively to CIE standard illuminants A, B, C, and D_{65}, but these illuminants are best used with a qualified matrix **R** of rank six.

Nevertheless, I advise strongly against the use of an achromatic axis other than the equal-energy stimulus employed in the standard color-matching experiment. This canonical configuration has elegant algebraic and geometric properties that are lost with a different achromatic axis. Furthermore, Burns, Cohen, and Kuznetsov (1989) have demonstrated that the illuminant is best incorporated in a matrix **R** of rank six rather than three. By our view, there is only one canonical matrix **R**.

Configuration Referenced to Triplet of Orthogonal Spectral Vectors

In 1953–1954, long before the advent of matrix-**R** operations, MacAdam (1953, 1954, 1975) introduced "orthogonal and normalized functions," as they were then termed in colorimetry. Although MacAdam did not discuss his procedure explicitly, I surmise that he used a rotary calculator to generate the functions by the Gram-Schmidt orthogonalization. Given one orthonormal set, he transformed to other sets (e.g., by the well-known procedure described in earlier chapters here) in search of orthonormal functions that corresponded to spectral color stimuli. MacAdam thus endeavored to answer the question: Are there three orthogonal vectors that lie exactly on the spectral envelope?

Perhaps due to the computational limitations of the time, MacAdam was unable to find such vectors. For example, he found that 557 nm and 654 nm were orthogonal, but with one remaining degree of freedom, the third orthogonal vector was at 463 nm at about 94% purity on the CIE chromaticity diagram. By my calculations, this third orthogonal vector is about 12 degrees from the spectral envelope and this, of course, is unacceptable.

In 1985, Cohen and Kappauf (1985) turned their attention to this problem and, by Kappauf's logic applied to matrix **R**, demonstrated the existence of three mutually orthonormal spectral color stimuli to within extremely close tolerances, for both the 1931 observer and the 1964 observer respectively:

For the 1964 data:	455 nm	513 nm	584 nm
For the 1931 data:	457 nm	519 nm	583 nm

First, we transformed a 300-row **A** matrix (1964 color-matching functions 401 nm to 700 nm, CIE X,Y,Z) to these primaries using the methods described above. As expected, the rows of 455 nm, 513 nm, and 584 nm formed a 3×3 identity matrix. If these functions are orthogonal, we expected that the corresponding **M** and $\mathbf{M}^{-1}$ matrices ought to be diagonal matrices to within very close tolerance. Here is the **M** matrix:

$$\begin{bmatrix} 46.8111 & 0.0191 & 0.0192 \\ 0.0191 & 81.2359 & 0.1567 \\ 0.0192 & 0.1567 & 69.0238 \end{bmatrix}.$$

Here is the $\mathbf{M}^{-1}$ matrix:

$$\begin{bmatrix} 0.0214 & 0.0000 & 0.0000 \\ 0.0000 & 0.0123 & 0.0000 \\ 0.0000 & 0.0000 & 0.0145 \end{bmatrix}.$$

This last matrix is a subsection of matrix **R**, as are all $\mathbf{M}^{-1}$ matrices with spectral primaries.

Then we computed the associated orthonormal **F** matrix by multiplying each column of matrix **A** by the corresponding row vector length (the square root of the diagonal of the $\mathbf{M}^{-1}$ matrix). **F′F** should give an identity matrix. Here is the identity matrix so obtained:

$$\begin{bmatrix} 1.0000 & 0.0003 & 0.0003 \\ 0.0003 & 1.0000 & 0.0021 \\ 0.0003 & 0.0021 & 1.0000 \end{bmatrix}$$

Closely similar matrices were obtained for the CIE 1931 standard observer.

For the 1964 observer, the spectral orthogonal reference axes (vectors) are illustrated in Figure 24A. Although the view is distorted by perspective, the reference axes are mutally orthogonal and lie precisely on the envelope of the fundamental color space. The scalar fundamentals, the columns of matrix **F**, are shown in Figure 24B; note that all orthonormal functions have mutual crossovers at zero. Kappauf affectionately gave the name *golden vectors* to this set of orthonormal vectors.

Figures 24C–E illustrate these golden functions plotted separately, to illustrate that an **F** matrix is actually both an **A** matrix and an **E** matrix with different scales; this has already been discussed. In each case, the functions are identical, but the y-axis has been multiplied by different coefficients. *C*, *D*, and *E* are the scalar orthonormal functions in *B*, but plotted separately with a triple of scales to the left of each function. The first scale is the orthonormal metric and identical to the metric in *B*. The second scale converts these orthonormal functions to color-matching functions by dividing each value in the orthonormal metric by the root of the respective vector length. The third scale converts these orthonormal functions to fundamentals by multiplying each value in the orthonormal metric by the square of the respective vector length. When the vector lengths of orthonormal functions are known, the orthonormal functions may be converted to color-matching functions or fundamentals. The difference is merely a scaling factor, but note that the scaling factor is different for each orthonormal function, because the respective vector lengths are different.

How were these orthonormal functions located to this precision, where others had failed? We had a strategy. We first assumed that the triplet did exist. If true, then we expected that the first function would be in the shorter wavelengths (400–459 nm), the second function would be in the mid-range wavelengths (500–520 nm), and the third function in the longer wavelengths (570–590 nm). We then generated a 300 × 300 matrix **R** with 90,000 entries; because of computer limitations at that time, the **R** matrix was computed 27 columns at a time, by the equation $\mathbf{AM}_e = \mathbf{E}$.

Now, the entries in matrix **R** are scalar products, and a zero entry signifies orthogonal fundamentals—except in the trivial case where one of the fundamental vectors is zero. In any column of matrix **R**, the crossover from plus to minus was presumed to represent a zero scalar product on the side with the smallest absolute value.

The orthogonal triplet was found by plotting the three functions in Figure 25. Graph A plots the locus of paired orthogonal fundamentals—one from the shorter range and one from the middle range; for example, 420 nm is closely

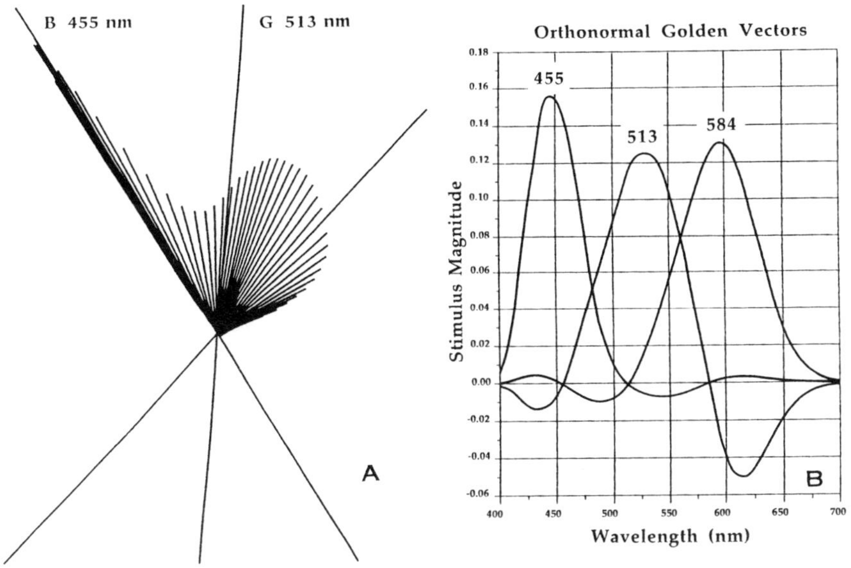

Figure 24. "Golden vectors" and scalar orthonormal functions. *Continued on next page.*

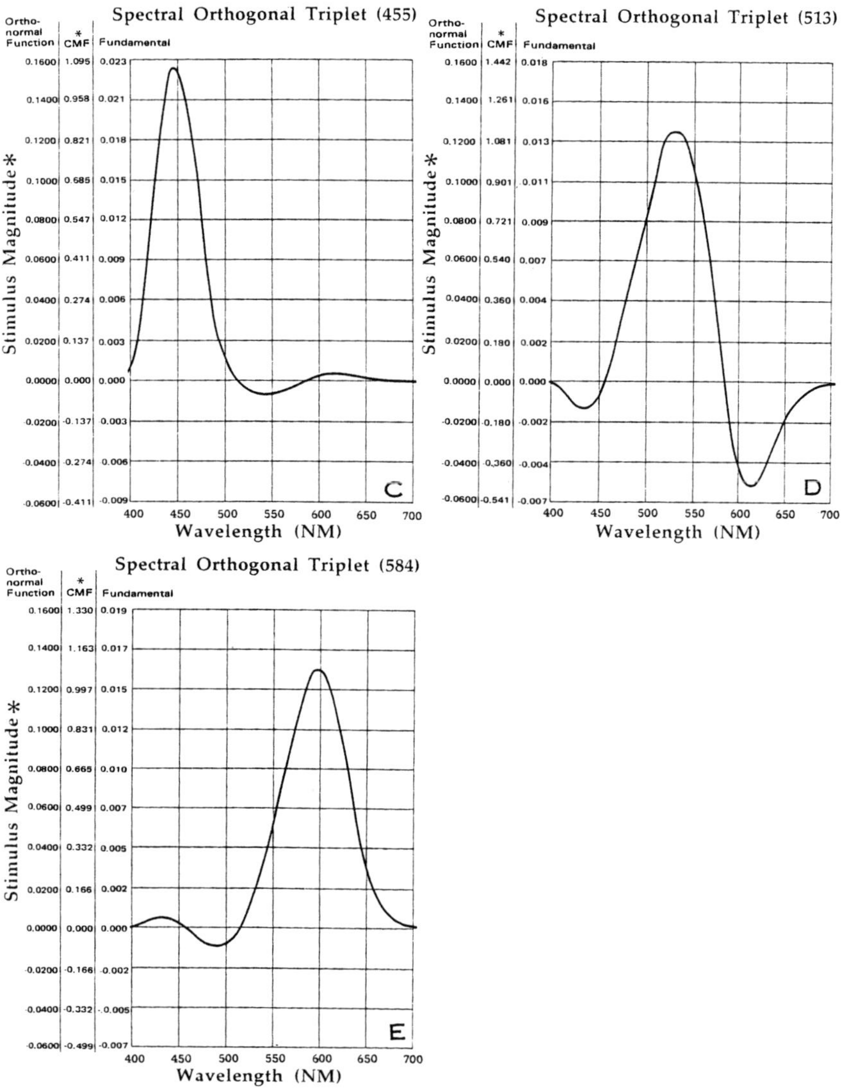

Figure 24. Continued.

orthogonal to 500 nm. Next, we sought two wavelengths, one from the middle range and one from the longer range that were both simultaneously orthogonal to one or another of the wavelengths in the shorter range *and* to each other. Graph B plots the locus of paired orthogonal wavelengths—one already specified by graph A, and the other from the longer wavelengths. Graph C

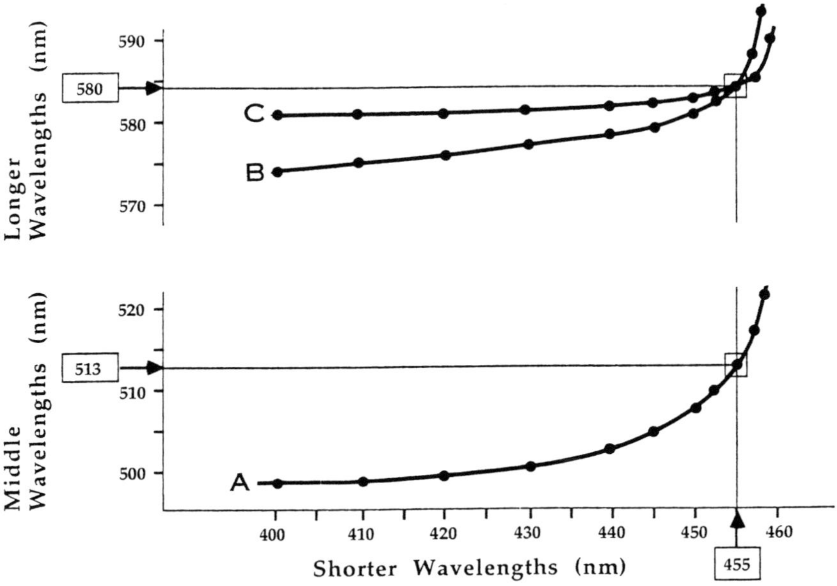

Figure 25. Search for the "golden vectors."

plots the locus of paired orthogonal wavelengths—one from the shorter wavelengths and the other corresponding to the middle wavelengths, referring to graph A, which was found to be orthogonal to each of the shorter wavelengths. Immediately, these functions indicate triplets of potentially mutually orthogonal spectral vectors.

For example, note that graphs B and C cross at 584 nm at the longer wavelengths. This means that 584 nm is orthogonal to both 455 nm and 513 nm; therefore this is an orthogonal triplet.

13

Complementary and Orthocomplementary Color Stimuli in Fundamental Color Space

Complementary stimuli depend on the definition of a stimulus that evokes a sensation of a given magnitude, described by the word *white*. Then, complementary stimuli are defined as either (1) two color stimuli whose mixture evokes the specified sensation, but at different magnitude or (2) two color stimuli whose mixture evokes the specified white sensation at the specified magnitude. The first definition is weak; the second is far more rigorous. Colorimetrists usually use the first, stating that wavelengths are complementary when their mixture gives a white of any magnitude. I usually use the second, because this analysis is permitted by matrix **R** operations.

Complementary sensations are often defined subjectively, as, for example, the after-image of a sensation evoked by a "blue" stimulus (see Wilson & Brocklebank, 1955). I am not concerned here with this type of definition depending on "accidental colors," although these after-images and induced color sensations play a significant role in the evolution of this subject, as I shall demonstrate.

The designator *complementary* in colorimetry derives from the Latin *complementum*, meaning "that which fills up or completes." The *Oxford English Dictionary* gives this definition: "Something, which when added, makes up a whole; each of two parts which mutually complete each other, supplying each other's deficiencies." The word has long been used in Euclidean geometry with respect to angles—an angle plus its complement equals 90 degrees. The word is also employed in astronomy and in music. Its use in colorimetry, however, is a fascination. The word was employed correctly more than fifty years before colorimetrists provided extensive, empirical demonstrations of complementaries; indeed, many colorimetrists doubted the existence of complementary spectral stimuli until well into the nineteenth century.

Colorimetrists have agreed that complementary color stimuli occur in pairs, and when additively compounded in proper proportions, they evoke a color sensation described by the word *white*. The specific white sensation is arbitrary; in this text, white is the sensation evoked by an equal-energy stimulus. Complementary color stimuli may be both monochromatic or spectral, one monochromatic and one heterochromatic, or both heterochromatic.

We may comprehend some aspects of complementary color stimuli in physical terms. Given the radiometric function of any color stimulus, its complement is easily computed. As noted in Hardy and MacAdam (1936), a complementary stimulus of a given radiometric function can be achieved by subtracting the radiometric function from an equal-energy (flat) radiometric function. This follows from the fact that a horizontal line at the upper limit represents a stimulus spanning the visible spectrum, with every monochromatic passband at equal energy. Indeed, this flat radiometric function may be split in any manner, horizontally or vertically or in any combination, and the two components must be complementary. Furthermore, a radiometric function of any metamer to this same white, even where the radiometric function is highly variable and ill behaved, may be split similarly, and the two components must be complementary still again. This is illustrated in Figure 13, along with complementary metameric suites.

Complementary spectral color stimuli are an astonishing aspect of colorimetry. A "white" sensation may be evoked by a stimulus compounded of all spectral color stimuli spanning the visible spectrum, each stimulus of equal physical magnitude (power); the spectrum may be split at any point, and the two remaining heterochromatic components are always complementary. Simultaneously, the identical "white" sensation may be evoked by a stimulus compounded of only *two* spectral color stimuli, although almost always of unequal magnitude.

The concept of complementarism is a special case of the broader concept of metamerism; two different complementary pairs evoke the same white sensation, although the compounded stimuli comprise vastly different radiometric distributions. Complementary colors have been investigated empirically and experimentally by Young (1807), Maxwell (1860), Helmholtz (1855), von Frey and von Kries (1881), König (1894, 1896), Grünberg (1904), Priest (1921–1922), Sinden (1923), Genberg (1933), Kirschmann (1917), Judd (1924), Warburton (1936), Schrödinger (1920b), Godlove (1930), MacAdam (1938), and Pridmore (1978, 1980, 1988).

Complementary colors are closely allied with the so-called *accidental colors*, that is, with colored shadows, simultaneous color contrast, and colored after-images. In every case, an adequate stimulus, evoking a "real" color sensation, may induce an "unreal" color sensation lacking an adequate color stimulus. In every case, the "real" and "unreal" color sensations are related by complementarism. On this basis, color theorists, such as Chevreul (1839) and Field (1835, 1845), have long proposed that complementary color stimuli evoke the most harmonious or consonant color sensations, while other pairs of color stimuli may evoke cacaphonous or dissonant color sensations. Color order systems, like those of Munsell and Ostwald, have exploited this precept by locating complementary pairs in linear opposition, on either side of the stimulus evoking white or gray.

Here I explore the significance of complementary color stimuli as revealed by matrix-**R** operations and the canonical representation of the fundamental color space. In fundamental color space, complementary color stimuli are represented by vectors coplanar with the "white" vector or axis; stated differently, all three vectors lie on the same plane. However, fundamental color space also reveals *orthocomplementary* pairs, where a color stimulus (vector) and the white axis define one plane, and the orthocomplementary vector and the white axis define another plane orthogonal or perpendicular to the first. The intersection of the planes is coincident with the white vector. As I shall demonstrate, some spectral color stimuli have only one orthocomplementary, while other spectral color stimuli have two orthocomplementaries that, of course, must be complementary to each other. In a mathematical sense, orthocomplementaries are the obverse of complementaries. If we accept the contention that consonance describes color sensations related to complementary stimuli, perhaps it may be expected that dissonance describes color sensations related to orthocomplementary stimuli.

Chronology of Complementarism

The history of complementary color stimuli accurately reflects the history of colorimetry and is essential to my arguments.

In his first paper on prismatic colors, Isaac Newton (1671–1672) was surely aware of additive color mixture and the range of differential color sensations, but he was not aware of metamerism; he was unable to evoke precisely the same color sensation from two different spectral distributions. Newton wrote:

> There are therefore two sorts of Colours. The one original and simple, the other compounded of these. The original or primary [spectral] colours are, *Red, Yellow*, Green and a Violet purple, together with Orange and Indigo, Blew, and an indefinite variety of Intermediate gradations.
>
> The same colours in *Specie* [but not identical in every respect] with these primary ones may be also produced by composition; For a mixture of *Yellow* and *Blew* makes *Green*; of *Red* and *Yellow* makes *Orange*, of *Orange* and *Yellowish green* makes *yellow*. And in general, if any two colours be mixed, which in the series of those generated by the Prisme, are not too far distant one from another, they by their mutual alloy compound that colour, which said series appeareth in the mid-way between them. But those, which are situated at too great a distance, do not so. *Orange* and *Indigo* produce not the intermediate Green. . . . To the same degree of Refrangibility ever belongs the same colour, and to the same colour ever belongs the same refrangibility. . . .
>
> But the most surprising and wonderful composition was that of *Whiteness*. There is no one sort of rays alone can exhibit this. 'Tis ever compounded, and to its composition are requisite all the aforesaid primary Colours. . . . But to determine more absolutely, what light is, after what manner refracted, and by what modes of actions it produceth in our minds the Phantasms of Colours, is not so easie.

Indeed, not so easy!

Furthermore, Newton argued that additive color mixture (to use recent terminology) was dependent on the spectral composition of compounded color stimuli. In response to a published rebuke, or animadversion as it was then called, leveled by an "ingenious person from Paris," Christiaan Huygens (1673), Newton (1673) reiterated this position in the *Philosophical Transactions*:

> But yet there is an experiment or two mentioned in my [previous] letter. . . by which I have produced White out of two colors alone, and variously, as out of *Orange* and *full Blew*, and out of *Red* and *pale Blew*, and out of *Yellow* and *Violet*, as also out of other pairs of Intermediate colors. The most convenient Experiment for performing this, was that of casting the colors of the Prisme upon those of another, after a due manner. But what [the critic] can deduce from hence, I see not. For the two colors were compounded of all others, and so the resulting White (to speak properly,) was compounded of them all, and only decompounded of those two. For *instance*, the *Orange* was compounded of Red, Orange, Yellow, and some Green; and the *Blew*, of violet, full Blew, light

> Blew, and some green, with all their Intermediate degrees; and consequently the Orange and Blew together made an Aggregate of all colors to constitute the White. . . . This is so plain, that I conceive there can be no further scruple to them who know how to examine, whether a colour be simple or compounded, and of what colors it is compounded. . . . If therefore [the critic] would conclude anything, he must shew, how white may be produced out of two *Un-compounded* colors. . . . But I believe that there cannot be found a [successful] Experiment of that kind; because, as I remember, I once tryed, by gradual succession, the mixture of all pairs of Un-compounded colors; and, though some of them were pailer, and nearer to White, than other, yet none could be truly call'd White. But it being some years since this tryal was made, I remember not well the circumstances, and therefore recommend it to others to be tryed again.

Nevertheless, in his *Opticks* of 1704 (Book I, part 2), Newton (1704) attempted a prediction of the color sensation evoked by a compound stimulus, that is, evoked by a mixture of several different spectral color stimuli. Proposition 6, Problem II stated: "In a mixture of primary [spectral] colours, the quantity and quality of each being given, to know the Colour of the Compound." As an approximate solution, Newton proposed a two-dimensional planar model. In a circle with white at the center, Newton divided the circumference into seven "hue-definitive" arcs whose spectral stimuli evoked generally red, orange, yellow, green, blue, indigo, and violet color sensations, respectively. The arc lengths were presumed proportional to the seven diatonic musical intervals; Newton had conjectured (incorrectly) that as the musical harmonies of the Pythagoreans had permeated the musical sensations elicited by auditory stimuli, so also had the celestial harmonies permeated the color sensations elicited by the visible spectrum; this, of course, was an application of the views propounded by Johannes Kepler in his *Harmonices Mundi*.

Newton placed a small circle at the gravitational center (centroid) of each arc, the magnitude of the small circle proportional to the "amount" of each spectral color stimulus in the compound. The combined centroid of these seven centroids indicated approximately the psychological sensation evoked by the compound stimulus. A radius was drawn from the central white point through the compounded centroid. The intersection with the circumference roughly indicated hue, and distance from the center roughly indicated whiteness (chroma or saturation). "This [barocentric] rule," wrote Newton, "I conceive accurate enough for practice, but not mathematically accurate."

Although his color diagram indicated that complementary color stimuli should exist, Newton produced no experimental evidence. In his *Opticks* (Book 1, part 2), Newton gave a final assessment:

> For I could never yet by mixing only two primary [spectral] Colours produce a perfect white. Whether it can be compounded of a mixture of three taken at equal distances in the circumference I do not know, but of four or five I do not much question but it may. But these are curiosities of little or no moment to the understanding the Phenomena of nature. For in all whites produced by nature, there uses to be a mixture of all sorts of rays, and by consequence a composition of all Colours.

Newton's statements appear definitive and were reviewed by Alan E. Shapiro (1984). Nevertheless, a bitter debate developed between David L. MacAdam and Alan E. Shapiro (MacAdam & Shapiro, 1985). MacAdam asserted that Newton had, in fact, demonstrated that a stimulus evoking a white sensation could be compounded of three or even two spectral color stimuli. Shapiro (1980), having researched and written papers on Newton's color circle and Newton's evolving conception of white light, asserted that Newton had not made the demonstration and, furthermore, that Newton had not claimed to have done so.

Newton's planar diagram, so primitively conceived and printed in *Opticks*, is the great forebear of the invariant fundamental color space described in this book.

The word *complementary* was introduced into colorimetry by an enigmatic and provocative individual, Sir Benjamin Thompson or Count Rumford, an American born in Woburn, Massachusetts. Rumford was an active Tory during the initial stages of the American Revolution and then embraced English ideals, becoming an administrator in Great Britain, charged with governing the colonies under King George III. Rumford had a short military career fighting against the Americans, returned to England, and became one of the few Americans to be knighted by the Crown. He eventually resided in France, where he married the widow of Antoine Lavoisier, who had been guillotined during the French Revolution. He founded the Royal Institution of Great Britain to provide education for the poor, supplied funds for Rumford prizes still awarded by the Royal Society of London and by the American Academy of Arts and Sciences, and also funds for the Rumford professorship still active at Harvard University. He wrote several papers on the preparation of soup for the lower classes and on the exhilarating tastes of coffee.

Rumford is best remembered for his abiding interest in heat dissipated during the boring of cannon. His experiments in this area annihilated the popular concept of caloric, a hypothetical fluid whose presence in matter was presumed to determine its thermal state, that is, its temperature. Rumford also was interested in light and described and built the photometer that bears his name. He was also interested in color, first in colored shadows and then in the complementary nature of colored shadows.

After conveniently forgetting that he had read about colored shadows in Joseph Priestley's *History of Optics* (1772), Rumford (1870–1875) rediscovered the phenomenon. With respect to this subject, he addressed the Royal Society of London on February 20, 1794, and reported:

> Desirous of comparing the intensity of the light of a clear sky, by day, with that of a common wax candle, and letting the daylight from the north... fall upon a sheet of very fine white paper, I placed a burning candle in such a position that its rays fell upon the same paper.... [I interposed] a cylinder of wood, about half an inch in diameter, before the centre of the paper.... I was much surprised to find that the shadows projected by the cylinder upon the paper, instead of being merely shades, *without colours* [hue], as I expected to find them—that which, corresponding to the beam of daylight, was illuminated by the candle—was *yellow*; while the other, corresponding to the light of the candle,—and consequently illuminated by the light of the heavens,—was of the most beautiful *blue* that it is possible to imagine. This appearance, which was not only unexpected, but was really in itself in the highest degree striking and beautiful, I found upon repeated trials and after varying the experiment in every way I could think of, to be so perfectly permanent that it is absolutely impossible to produce two shadows at the same time from the same body, the one answering to a beam of daylight and the other to the light of a candle or lamp, without these shadows being colored, the one *yellow* and the other *blue*.
>
> ... In the course of these experiments, I observed that different shades of yellow, given to the daylight [by filters], produced very different and often quite unexpected effects: Thus one sheet of the yellow glass, interposed before the beam of daylight, changed the yellow shadow to a lively violet colour, and the blue shadow to a light green ...
>
> Having no other coloured glass at hand to push these particular inquiries farther, I now removed the candles, and opening two holes in the upper parts of the window-shutters of two neighboring windows, I let into the room, from above, two beams of light from different parts of the heavens; and, placing the instrument in such a manner that two distinct shadows were projected by the cylinder upon the paper, I was entertained by a succession of very amusing appearances.
>
> The shadows were tinged with an infinite variety of the most unexpected and often most beautiful colours, which continuously varying, sometimes slowly, and sometimes with inconceivable rapidity, absolutely fascinated the eyes, and commanding the most eager attention, afforded an enjoyment as new as it was bewitching.

> It was a windy day, with flying clouds, and it seems as if every cloud that passed brought with it another complete succession of varying hues and most harmonious tints.

Soon after the delivery of this communication on colored shadows, Rumford (B. Thompson, 1802, 1870–1875) composed a sequel, "Conjectures Respecting the Principles of the Harmony of Colours," that appeared originally only in his collected *Philosophical Papers*. Rumford wrote: "In the experiment of the coloured shadows, the colour exhibited by one of the shadows only is real, that of the other is *imaginary* [unreal], being an optical deception. . . . The *imaginary colour*, which may be said to be *called up in the mind* by the other *real colour*, does not, however, appear to be at all inferior to the real colour either in lustre or in the distinctness of its hue."

Rumford extended his research using pigments:

> The experiment was first made at an inn at Florence, in the year 1793; and in order that I might assure myself that my expectations had not deceived me, by imposing on my senses, I called two of my friends who happened to lodge at the house (Lord and Lady Palmerston [whose son was to become prime minister of England under Queen Victoria]) into the room, and without letting them into the secret simply asked them, with a feigned air of indifference, which of the two colours they saw in the centre of the circular piece of paper on the floor they thought the brightest.
>
> After looking at them for some time, and going around to view them from different sides, one of them answered: "I don't know which of them is brightest. The red is very bright, and so is the blue. But why do you ask the question?"
>
> When I told them there was no blue there, and what they took to be blue was merely a deception, they did not believe me; but they were much surprised, and convinced that what I told them was true, when they saw that on my removing the red slip its companion, which was left behind, instantly *faded* and *lost its colour*. . . .
>
> The two colours exhibited by the two shadows appear in all cases to harmonize in the most perfect manner, or, in other words, afford the most pleasant contrast to view. . . .
>
> By experiments of this kind, which might easily be made, ladies may choose ribbons to their gowns; or those who furnish rooms may arrange for their colours upon principles of the most perfect harmony and of the purest taste. . . .
>
> By a careful examination of the works of the great masters of the art of colouring, it will appear that they have frequently practiced upon these

> principles, although it is not likely that they were acquainted with the scientific foundation of their practice. . . .
>
> These two colors [exhibited by the shadows] are always such that, if they could be intimately mixed together, the result of that mixture would be *perfect whiteness*; and, as whiteness results from the mixture of all the different colours in certain proportions, the two colors may be considered as containing all the colours in their just proportions, and the colour of the one shadow may with propriety be said to be the *complement* of the other.

The complementarism problem was dormant essentially until the end of the nineteenth century, when the giant intellects of three men assessed the color problem to develop a mathematical solution: Hermann von Helmholtz, distinguished and acclaimed professor of physics in the University at Berlin; Hermann Günther Grassmann, unacclaimed and almost unnoticed high school instructor of mathematics and physics at the Gymnasium in Stettin, Pomerania (now Poland); and James Clerk Maxwell, distinguished and acclaimed professor of experimental physics at the University at Cambridge.

Helmholtz was a great luminary of modern science, having made contributions to physiological acoustics, non-Euclidean geometry, hydrodynamics, electrodynamics, epistemology, and an area with a new name, physiological optics. The third edition of his authoritative treatise *Handbuch der Physiologischen Optik* was translated by J. P. C. Southall for the Optical Society of America in 1924 (Helmholtz, 1856–1867).

Helmholtz (1852) made his first significant contribution to physiological optics on the theory of compound colors in 1852; therein, additive color mixture was finally disassociated from subtractive color mixture. Here is an English translation of this pivotal moment in the history of color science:

> It will . . . be observed, that my results on the action of prismatic colours differ materially from those obtained by the mixing of colored substances [pigments or powders]. In particular, that [the mixture of] yellow and blue [prismatic colors] do not furnish green, but at most a weak greenish white, contradicts in the most decided manner the experience of . . . painters during the last thousand years. . . . The substances used in painting, as all coloured bodies of regular structure . . . are transparent, or at least translucent. . . . Let us consider the case when we mix together powders of different colors, for example, yellow and blue. The blue particles which lie on the surface will give [reflect] blue light, the yellow which lie on the surface will give [reflect] yellow light; both together will combine to form white or greenish-white. It is quite otherwise, however, with the light that returns from greater depths. This [light] must pass

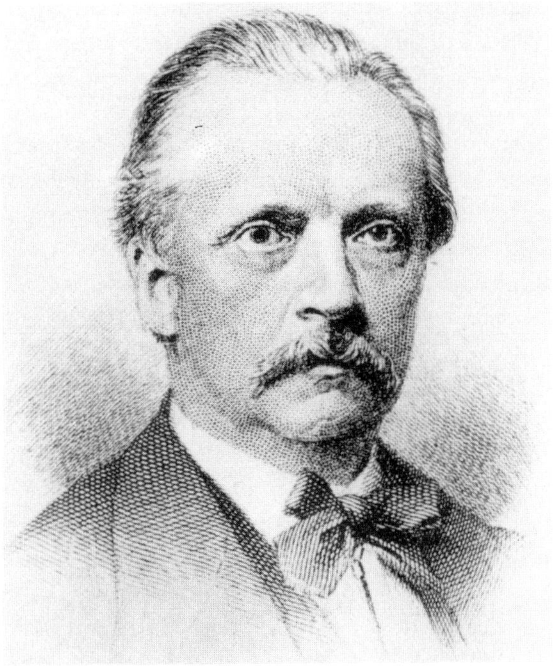

Hermann Ludwig Ferdinand von Helmholtz, 1821–1894

through [both] yellow and blue particles, and hence from a distance below the surface such light only will return as can penetrate both the yellow particles and the blue ones. Blue substances generally permit green, blue, and violet light to pass through them in sensible quantity; yellow, on the contrary permits red, yellow, and green to pass. Green, therefore, is the only light which will pass through both, and hence from the deeper layers of the mixed powder only green light can return. Now, as the quantity of the superficial portions of the powder is . . . generally much smaller than that which returns from the deeper layers, the consequence is, that the green of the latter is by far the most predominant, and thus determines the colour of the mixture. . . . The theory of pigmentary colours here presented is simply derived from the generally recognized laws of physics; it explains the phenomena, as far as I am able to see, completely; showing that the mixture of the substances and the combination of their colours are two processes altogether distinct, and hence that the results obtained from the former furnish no conclusion regarding the latter.

As we now know, the difference between additive and subtractive color mixture is closely allied to the phenomenon of complementarism, and indeed, modern printing inks—magenta, yellow, and cyan—are based on complemetary relationships.

With the matter of additive and subtractive color mixture resolved, Helmholtz attended to the foundations of trichromacy. Finally, he emphatically rejected Brewster's contention that the visible spectrum was composed of three superimposed continuous spectra and concluded tentatively that Thomas Young was correct in his final interpretation: that the color sensations evoked by additive color mixture depended on the internal channels of the color processing mechanism.

Therefore, Helmholtz chose to test this hypothesis experimentally. With apparatus vastly improved over the original Newtonian prisms, Helmholtz arranged the additive mixture of every pair of spectral color stimuli and, most important, the gradual intensity change of one or both members of the pair.

Despite the care in design and construction of his apparatus, Helmholtz reported extraordinary results and came to anomalous conclusions. He found yellow and blue (indigo) as the only pair of complementary spectral color stimuli, contrasting with Newton, who had found no pairs at all. From his artifactual results, Helmholtz concluded that an imitation of all spectral colors would require at least five primitive colors: red, yellow, green, blue, and violet. If restricted to three, the best primitive colors would be red, green, and violet—as in Young's final proposal.

Besides the complementaries yellow and blue, Helmholtz found that he could obtain white from an infinity of combinations. Presumably this was the first truly valid demonstration of metameric color stimuli. And so, Helmholtz emphatically rejected the physiological theory of Thomas Young:

> We must also abandon the theory of three primitive colours, which according to Thomas Young, are three fundamental qualities of sensation. If the sensation of yellow by the yellow rays of the spectrum were due to the fact that by them the sensations of red and green were simultaneously excited, and both working together produced yellow, exactly the same sensation must be excited by the simultaneous action of the red and green rays; nevertheless, by the latter we can never obtain so bright and vivid a yellow as that produced by the yellow rays.

In modern terminology, Helmholtz had found that a color sensation evoked by a spectral color stimulus cannot be matched by any combination of any other spectral color stimuli. This is true, of course, but the perplexity has been solved with negative tristimulus values.

With the exception of the specified differences between additive and subtractive color mixture, Helmholtz's theory of compounded color stimuli was a scientific calamity. The most honored name in color science was capable of appalling experimental error and had propounded an absurd color theory. Helmholtz, in fact, still lacked an essential construct.

With the advent of Newton's *Principia*, mathematicians realized that classical Euclidean geometry was essentially different from Newtonian geometry. For example, Bochner (1962) wrote:

> The Euclidean Space that underlies the *Principia* is not mathematically quite the same as the Euclidean Space that underlies Greek mathematics . . . from Thales to Appollonius. The geometry of the Greeks emphasized congruences and similarities between figures, that is, in analytical parlance, orthogonal and homothetic transformations of the underlying space. . . . The Euclidean space of the *Principia* continued to be all this, but it was also something new in addition. Several significant physical entities of the *Principia*, namely, velocities, momenta, and forces are, by mathematical structure, vectors, that is, elements of vector fields, and vectorial composition and decomposition of these entities constitute an innermost scheme of the entire theory.

Therefore, in the nineteenth century, the mathematical domain of vector analysis was founded, and its development still continues. There were precursors, however, as always. The addition of force vectors in parallelogram construction was known to Archimedes and also to Hero of Alexandria; this construction is true only by definition, as we are informed, and cannot be derived, say, from Newton's laws, although periodic attempts continue to permeate the literature. In 1827, August Ferdinand Möbius published *Der Barycentrische Calcul*, containing many vectorial constructs. Most important, Sir William Rowan Hamilton (1853), who invented the words *vector* and *scalar*, presented *Lectures on Quaternions* in 1853, a complex vector-like mathematics that James Clerk Maxwell despised. (Sir William Rowan Hamilton was Irish and not to be confused with his contemporary Sir William Hamilton, the Scottish metaphysician.) The title page of the quaternion volume identified Hamilton as the recipient of multitudinous honors, including fellow of the American Society of Arts and Sciences and member of the Royal Academy of St. Petersburgh. Despite Hamilton's prestige, except for use in the mathematics of rotating bodies and in "Clifford algebras" (the algebras of geometry), quaternions have virtually disappeared from modern mathematics and are not the foundation of contemporary linear algebra.

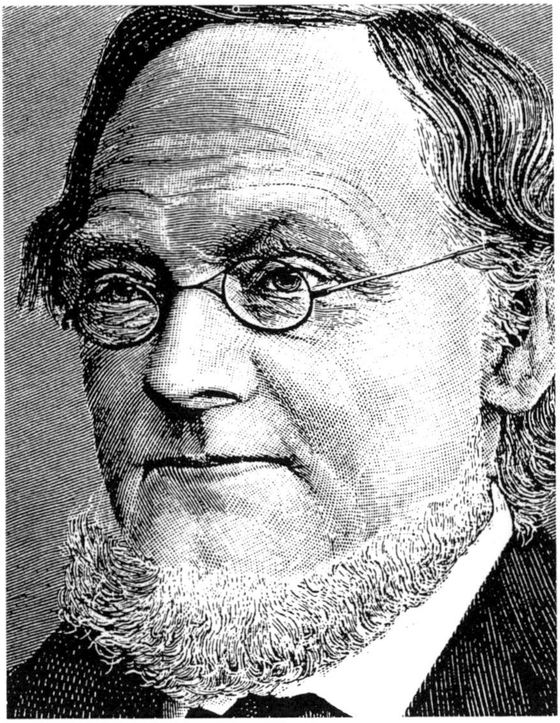

Hermann Günther Grassmann, 1809–1877

That understructure of vector analysis came, in fact, from the writings of the obscure Hermann Günther Grassmann, a virtually unknown Sanskrit scholar and political activist who had studied philology and theology (but not mathematics) for six months at the University of Berlin. In 1834, he returned to Stettin, where he remained a teacher—always well below the university level. In 1844, he wrote and self-published an extraordinary mathematical treatise on so-called extensive quantities with a 32-word title, usually shortened to *Die Lineale Ausdehnungslehre* (1894–1911). The title page identified Grassmann only as a high school teacher in Stettin. Grassmann described his work:

> It had for a long time been evident to me that geometry can in no way be viewed, like arithmetic or combination theory, as a [known] branch of mathematics; instead, geometry relates to something already given in nature, namely, space. I also had realized that there must be a branch of mathematics which yields, in a perfectly abstract way, laws similar to those in geometry, which appears bound to space. . . . Indeed, this new analysis, developed without the assumption of any principles established

> outside of its own domain and proceeding purely by abstraction, was itself this science.
>
> The essential advantages were [that] all principles which express views of space are entirely omitted and . . . the limitation to three dimensions is [also] entirely omitted.

Grassmann's *Die Lineale Ausdehnungslehre* was difficult to read and at the time almost impossible to comprehend; a single review appeared in a generally approved mathematical journal and that written by Grassmann himself—probably the only author ever to review his own work. Few copies of the *Ausdehnungslehre* were sold, and Grassmann sent most as gifts to recognized mathematicians and received one acknowledgment.

Despite the failure of *Ausdehnungslehre*, Grassmann remained convinced that his contribution had its own destiny and self-published a second edition in 1862; this also failed. Later generations of mathematicians have recognized the power of the *Ausdehnungslehre*. After Grassmann's death, the book was reprinted several times and may be purchased today from commercial publishers. Despite its early rejection and repudiation, the *Ausdehnungslehre* in fact harbored the basis of modern colorimetry, which Grassmann himself realized immediately.

In 1853, the inconsequential Grassmann criticized the 1852 color theory of imperious Helmholtz:

> Helmholtz published a series of observations . . . from which he comes to the conclusion, that the theory of compound colours, universally admitted since the time of Newton, is erroneous in its most essential points; and that there are properly only two prismatic colours, yellow and indigo, which when mixed, give white. It may consequentially not be regarded as superfluous to show that the Newtonian theory of compound colours is correct to a certain point, and especially that every colour has its complementary colour, with which when mixed gives white, is found upon mathematically incontestable facts so that this opinion must be regarded as one of the most tenable in physics. I will then show that the *positive* observations of Helmholtz, instead of tending to the subversion of this theory, serve rather partly to prove and partly to complete it.

Grassmann's analysis is based entirely on logic—"mere mathematical manipulation" in the words of several critics uneasy with this level of advanced reasoning. In fact, mere mathematical manipulation is the basis of great scientific contributions, including Copernicus's heliocentric theory of planetary motion and Einstein's theory of relativity. So merely by thinking about it, Grassmann made one of the most important contributions to color theory—perhaps *the* most important.

In his communication, Grassmann noted that every color sensation has three degrees of freedom: tint (hue), intensity (brightness), and achromatic content (saturation). He further observed that language provides a vocabulary to describe these variables. Furthermore, Grassmann observed that color sensations form a closed continuous manifold, because it was always possible to pass from one color sensation to another color sensation by way of uninterrupted intervening color sensations; no discontinuity had ever been observed in the color manifold. This continuity, Grassmann argued, is observed in the gradual transition between spectral colors and even through purples, the sensations connecting extreme red and extreme violet. If one is given color stimulus A_1 on the spectral continuum, the gradual addition of white stimulus W and the gradual reduction of A_1 gradually diminish the sensation evoked by A_1 until it is replaced by the white sensation. Therefore, every color stimulus can be gradually and continuously changed to a white stimulus.

If given the A_1 stimulus on the spectral locus and gradually change the evoked sensation to that evoked by the A_2 stimulus, also on the spectral locus, we may pass along the periphery of the spectral locus in either one direction (positive) or the other direction (negative). If A_1 and A_2 are of sufficient distance, we may pass from A_1 to A_2 *through* the configuration. We may add white to A_1 as it is diminished, and we may add white to A_2 as it is diminished, so that both stimuli evoke the same white sensation. It follows, because the color manifold is closed, that for every A_1, there must exist a corresponding opposite, A_3, so that adding A_3 to A_1 brings a continuous sensation change from A_1 to A_3 that passes through white. The color stimulus A_3 must always exist for color stimulus A_1 and is said to be its complementary; similarly the color stimulus A_1 must always exist for color stimulus A_3 and is said to be its complementary. Using data from Newton and Helmholtz, Grassmann derived a small table of complementary color stimuli; green, he said, was complementary to purple, and yellowish-green complementary to violet.

Grassmann believed that the purples were part of the visible spectrum, the transitory stimulus between extreme red and extreme violet. He was wrong, of course; purple is not a spectral color stimulus, but rather the mixture of extreme reds and extreme violets. Nevertheless, the purples act as part of the continuum of spectral colors in the color manifold, and the nonspectral purples are complementary to the spectral greens.

In this same paper, Grassmann wrote, "In order to introduce the main principle of compounded colours, I add a third assumption . . . that two colours, each of which has a constant tint and a constant intensity of the intermixed white, also give constant mixed colours, no matter of what homogeneous colours they may be composed." In modern terminology, Grassmann states that the additive mixture of compound lights depends on the evoked sensations of the component stimuli, not on their radiometric composition. Thus, metameric color stimuli are equivalent in additive color mixture. Again, Grassmann proved this most basic theorem by "mere mathematical manipulation."

Then Grassmann proposed a new mathematics to predict additive compounding of colors:

> Suppose the intensities of the two colours to be mixed to be signified by the length of the lines representing them, so that if a colour has the tone *a*, and its intensity is in the same proportion to that of *a* as *m* to 1, then the colour may be represented by a line having the same direction as *a*, but *m* times its length. [The same for the other color.] Having represented in this manner the two colours geometrically, let us construct from these lines the *geometrical sum*, that is, the diagonal of the parallelogram which has the two lines for its sides, and assume that this sum or diagonal shall represent the colour of the mixture, its direction showing the tint, and its length the intensity of the colour.
>
> This done, the tint and intensity of any mixture of colours may be found by simple construction. Thus it is only necessary to determine the lines which represent the tint and intensity of the mixed colours, and then to add these geometrically, that is to *compound them as forces*, and the geometrical sum (the resultants of the forces) represents the tint and intensity of the mixture [italics added].

Again, Grassmann proved this powerful theorem by "mere mathematical manipulation."

Note carefully that *m*, in Grassmann's discussion, is a proportion and corresponds to a contemporary tristimulus value. Proportions are dimensionless numbers that multiply the vector length, and then the vectors are added geometrically. This is precisely the procedure used in matrix **R**. This is not the procedure of the CIE, where the tristimulus space, for example, is based on the multipliers themselves and not on the multiplied vectors.

Grassmann's paper had a salutary effect on Helmholtz (1855), who repeated his experiments with an alternative method suggested by Jean Foucault and confirmed experimentally that every color stimulus did indeed have a corresponding complementary, except the greens whose complementaries were the nonspectral compounded purples. Helmholtz graciously acknowledged his indebtedness to Grassmann.

I acknowledge my own intellectual debt to Grassmann. He stated that additive color mixture is dependent on color sensations evoked by compounded colored stimuli. Grassmann introduced vector algebra to color science, successfully solved the problem of complementaries, and, most important, invented the so-called inner and outer products that are so important in contemporary linear algebra. The inner or scalar product is the basis of matrix-**R** operations, and indeed, matrix **R** is nothing more than an ordered collection of scalar products between each pair of vectors representing spectral color stimuli. Grassmann never comprehended the connection of his scalar products to color theory. Shortly before his death, still largely unappreciated, Grassmann (1877) wrote a second paper on color. (For a biography of Grassmann, see Schlegel, 1878, and Engel, 1911. For a sterling analysis of the origin of Grassmann's *Ausdehnungslehre*, see Lewis, 1977. For a sterling analysis of the history of vector analysis, see Crowe, 1967.)

In 1868, Helmholtz (1873) delivered a series of popular lectures in Frankfurt and Heidelberg and, in an embarrassing moment in the history of science, sought to rehabilitate Sir Thomas Young:

> The theory of colours, with all these marvelous and complicated relations, was a riddle which Goethe in vain tried to solve; nor were we physicists and physiologists more successful. I include myself in this number; for I had long toiled at the task, without getting any nearer my object, until I at last discovered that a wonderfully simple solution had been discovered at the beginning of this century, and had been in print ever since for any one to read. This solution was found and published by . . . Thomas Young. . . . He was one of the most acute men who ever lived, but had the misfortune to be too far ahead of his contemporaries. . . . Thus a mass of important thought lay buried and forgotten, until a later generation by slow degrees arrived at the rediscovery of his discoveries, and came to appreciate the force of his arguments and the accuracy of his conclusions. . . .
>
> Dr. Young supposes that there are in the eye three kinds of nerve-fibers, the first of which when irritated in any way, produces the sensation of red, the second the sensation of green, and the third that of violet. He further assumes that the first are excited most strongly by the waves of the greatest length, the second . . . by waves of middle length; [and the third] . . . only by waves of the shortest vibrations. . . . It will be seen that this hypothesis is nothing more than a further extension of Johannes Müller's law of special sensations.

Compare this assessment of Young's place in color science with Helmholtz's prior assessment.

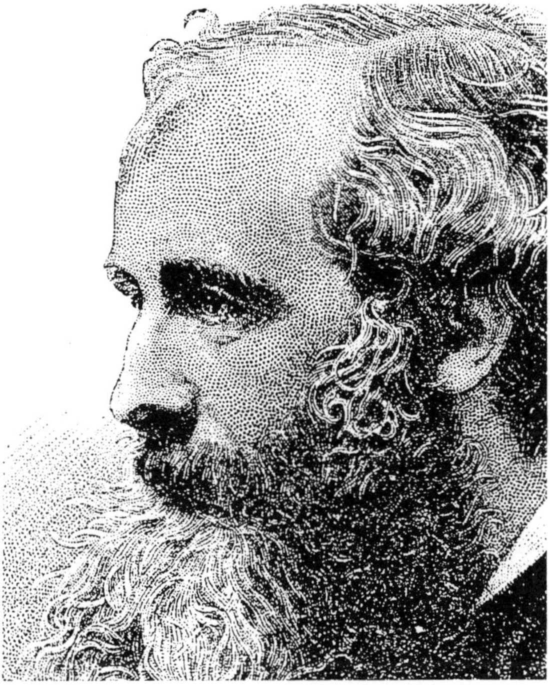

James Clerk Maxwell, 1831–1879

The reference is to Johannes Müller's law of the specific energies of nerves. The law states that the same stimulus acting on different nerves gives rise to different sensations in the sensorium, but different stimuli acting on the same nerve give rise to the same sensation in the sensorium. Stated differently, the law states that the sensation depends only on the nerve stimulated, and not on the properties of the stimulus. (See Müller, 1826, and Boring, 1929.)

On the occasion of James Clerk Maxwell's death in 1879 at age 48, W. D. Niven (1879) wrote a commemorative memoir in *Proceedings of the Royal Society (London)*: "It is seldom that the faculties of invention and exposition, the attachment to physical science, and capability of developing it mathematically, to describe the more delicate features of the works resulting from this combination, every one of which is stamped with the subtle but unmistakable impress of genius."

In contrast to Grassmann, the scientific community paid extraordinary homage to Maxwell during his lifetime. He was the first Cavendish Professor of Experimental Physics at Cambridge University. He received honorary degrees from the University of Edinburgh, Oxford University, and, with the Volta Medal, another from the University of Pavia. He held membership in the Royal Society of London and in numerous other societies, including the American Academy of Arts and Sciences, the Royal Society of Sciences at Göttingen, the New York Academy of Science, and the Imperial Academy of Sciences in Vienna. (Biographies of Maxwell include Campbell & Garnett, 1882, and Everitt, 1975.) Maxwell had many interests, including elastic solids, pure geometry, Saturn's rings, Faraday's lines of force, molecular physics, the zootrope (a primitive apparatus for displaying animated pictures), the stereoscope, and especially color vision. The colorimetric language itself records our debt to Maxwell; we have the Maxwellian view, the Maxwell disk, the Maxwell triangle, and the Maxwell spot.

On March 24, 1871, Maxwell delivered a lecture on color vision to the Royal Institution of London. He recorded his final appreciation of Thomas Young:

> All vision is colour vision, for it is only by observing differences of colour that we distinguish the forms of objects. I include differences of brightness or shade among differences of colour.
>
> It was in [this same] Royal Institution, about the beginning of this century, that Thomas Young made the first distinct announcement of that doctrine of the vision of colours which I propose to illustrate. We may state it thus:—We are capable of feeling three different colour-sensations. Light of different kinds excites these sensations in different proportions, and it is by the different combinations of these three primary sensations that all the varieties of visible colour are produced. In this statement there is one word on which we must fix our attention. That word is Sensation [*sic*!]. It seems almost a truism to say that colour is a sensation, and yet Young, by honestly recognising this elementary truth, established the first consistent theory of colour. So far as I know, Thomas Young was the first who, starting from the well-known fact that there are three primary colours, sought for the explanation of this fact, not in the nature of light, but in the constitution of man. Even of those who have written on colour since the time of Young, some have supposed that they ought to study the properties of pigments, and others, that they ought to analyze the rays of the light. They have sought for a knowledge of colour by examining something in external nature—something out of ourselves.
>
> Now, if the sensation which we call colour has any laws, it must be something in our own nature which determines the form of these laws. . . .
>
> Before the time of Newton, white was supposed to be of all known things the purest. When light appears coloured, it was supposed to have

> become contaminated by coming into contact with gross bodies. We may still think white light the emblem of purity, though Newton has taught us that its purity does not consist in simplicity. . . .
>
> All light, as Newton proved, is composed of these rays taken in different proportions. Objects which we call coloured when illuminated by white light, make a selection of these rays, and our eyes receive from them only a part of the light which falls on them. . . . This, then, is the optical explanation of the colours of bodies when illuminated with white light. They separate the white light into its component parts, absorbing some and scattering [reflecting] others.

Maxwell is correct essentially, but he suggests that the visual system processes the totality of the light reflected into the visual system. This is virtually untrue, even for monochromatic stimuli. Matrix-**R** operations demonstrate that the visual system in general processes only a portion of the color stimulus entering the visual system.

In his career, Maxwell (1854–1855) developed several apparatuses for the study of color mixture. The first was a modified rotating disk with colored sectors, called a *color top* or *teetotum*. Maxwell's color top had both an outer and an inner ring, the circumference of each delineated into an even 100 segments. (He might have used degrees.) The proportions of the colored sectors on the outer ring were changed under rotation, until they matched different proportions of different color stimuli on the inner ring. In his later experiments, for example, Maxwell was interested in the proportions of red (vermilion), blue (ultramarine), and green (emerald green) in the outer ring required to match a gray comprising black (B) and white (snow-white) in the inner ring. On the basis of 100 segments along the circumference of each ring, Maxwell had a color equivalence of sorts, where the coefficients were proportions:

$$\underset{\text{(outer ring)}}{.37\ \mathrm{V} + .27\ \mathrm{U} + .36\ \mathrm{EG}} = \underset{\text{(inner ring)}}{.28\ \mathrm{SW} + .72\ \mathrm{B}}. \tag{13.1}$$

Note that the sum of the coefficients in the outer ring and the inner ring totals unity.

Maxwell in fact was investigating complementary color stimuli. In the equation above, the sum of any two primaries is complementary to the third. For example, .37 V + .27 U is complementary to .36 EG, since the addition of these two components provides a stimulus that evokes a white sensation of lower magnitude—gray.

In this same paper, Maxwell anticipated the metameric blacks or residuals developed later in this text (see Cohen, 1988). Maxwell moved the right side of the colorimetric equivalence to the left:

$$54\ \mathrm{V} + 14\ \mathrm{G} + 32\ \mathrm{Y} - 32\ \mathrm{W} - 68\ \mathrm{B} = 0. \qquad (13.3)$$

The zero signified a theoretical black sensation, the first color-matching equivalence so written. This was the great ancestor of numerous similar equations written by Günter Wyszecki to generate metameric blacks in the 1960s.

These equivalences were not mathematical equations in any sense. The specified stimuli on the left evoke a sensation identical to the specified stimuli on the right. Nevertheless, the left side of the psychological equivalence does not equal the right side, physically or mathematically. Furthermore, the stimuli are not specified but merely named; the names have only subjective meaning and have no counterpart in modern mathematics.

From 1852 to 1856, Maxwell (1860) developed and refined his second colorimetric apparatus, the so-called color box; this was an ambitious apparatus for color matching, where prisms produced near-monochromatic stimuli. Maxwell had a scale and color names for wavelengths spanning the spectrum. His color-matching equations, however, matched white, and not another monchromatic stimulus as practiced by modern colorimetry. Spectral stimuli were not matched directly. Nevertheless, Maxwell did compute two triplets of functions—one labeled J (for James) and the other labeled K (for Katherine Dewar Maxwell, his wife). This was the first attempt to record systematically color-matching functions, and after some postexperimental data manipulation, Judd (1964) found a tolerable linear transformation of Maxwell's data to the standard CIE 1931 color-matching functions.

I note, however, that Maxwell was working with complementary color stimuli. Here is the mean equation of four sets of observations made by the same observer:

$$6.4\ \text{scarlet} + 25.2\ \text{yellow-green} + 31.3\ \text{blue} = \text{white}. \qquad (13.4)$$

Again, the sum of any two primaries is complementary to the third; thus, 6.4 scarlet + 31.3 blue is complementary to 25.2 yellow-green. (Again, these are not valid mathematical equations.)

In 1860, Maxwell was awarded the Rumford Medal by the Royal Society for his researches on the composition of colors; that paper is the single most important precursor to the newly developed matrix-**R** operations. In this award-winning paper, Maxwell disclosed a rarely referenced geometric construction for the representation of color sensations, similar to that already developed by Grassmann. Maxwell wrote: "It appears therefore that if colours are represented in quantity and quality by the magnitude and direction of straight lines [vectors], the rule for the composition of colours is identical with that for the composition of forces in mechanics. This analogy has been well brought out by Professor Grassmann." Matrix-**R** operations are the fulfillment of this statement by Maxwell more than 130 years ago.

As both Grassmann and Maxwell had observed, the elements of vector analysis are actually contained in the two-dimensional models of Newton. I call attention to two monographs, one by Preyer (1877) and the other by Hering (1887). Both attempted to use vectors and vector addition directly, but these researchers lacked the projection operator, the essential construct that has made matrix-**R** operations possible.

Complementarism and Orthocomplementarism as Disclosed by Matrix-R Operations

Projections onto the Chromatic Plane

Figure 14B demonstrates the canonical representation of fundamental color space. In that figure, the first EE axis (the achromatic axis) is orthogonal to the plane defined by the second and third axes orthogonal to each other (on the chromatic plane). Each spectral color stimulus is represented by a vector on the envelope of the configuration defined by a single matrix **F**. Every spectral vector projects onto the chromatic plane. Those projections are illustrated in Figure 26. (See also Figure 15D.)

These projections give insight into a salient property of fundamental color space. Figure 26A shows the projections of 300 spectral vector terminus points—401 nm to 700 nm—onto the chromatic plane (defined by orthogonal axes I and II). The spectral vectors cluster at the points of greatest curvature; note the marked increase of density at these points. This fact is dramatized by Figure 26B, where the reciprocal of the distance between adjacent vectors in *A* has been plotted. Figure 26C is an additional demonstration where every fifth vector is plotted. Clearly the projections of complementary vectors are at 180 degrees to each other, and the projections of orthocomplementary vectors are at 90 degrees to each other. Using an **F** matrix with 300 rows (1964 data, 401 nm to 700 nm) I have analytically computed the angles between each pair of vectors projected onto the chromatic plane. This is easily accomplished by matrix-**R** operations and matrix **F**. Given the second and third entries in two given rows of matrix **R**, I have the projections of the two vectors onto the two axes lying on the chromatic plane. Form a 2 × 2 matrix **J**. Then, $\mathbf{J}\,\mathbf{J}' = \mathbf{R}_V$, where $\mathbf{R}_V$ is a symmetrical subsection of matrix **R**, where the diagonals are the squares of the respective vector lengths and the side entry is the scalar product between the two given vectors. The side entry is divided by the product of the respective vector lengths to give the arc cosine of the intervening angle.

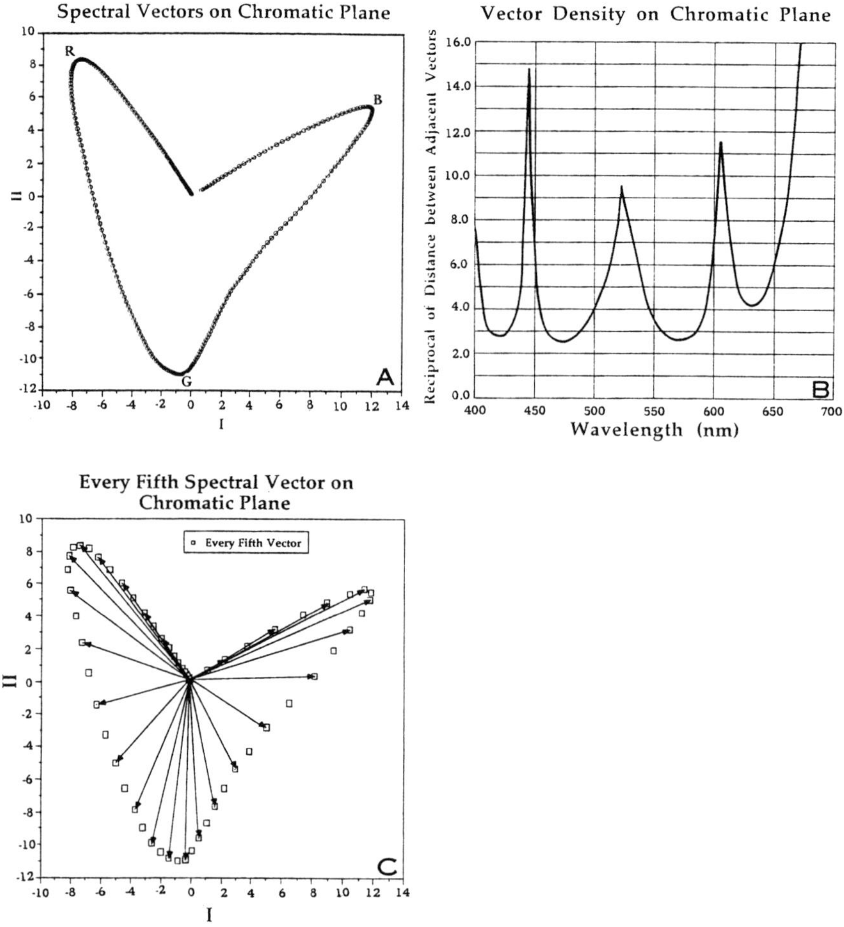

Figure 26. Spectral vector projections onto the chromatic plane.

The computer has generated these intervening angles for each pair of spectral vectors—90,000 pairs. For each spectral color stimulus, I selected the vector closest to 180 degrees as the complementary wavelength and the vector or vectors at 90 degrees as the orthocomplementary wavelength or wavelengths. Graphic plots of several spectral stimuli are shown in Figure 27. *A* plots the angle between the projection (onto the chromatic plane) of the vector fundamental for monochromatic 480 nm and the projections (onto the chromatic plane) of all other monochromatic vector fundamentals. *A* also plots the angle between the projection of the monochromatic vector fundamental 586 nm

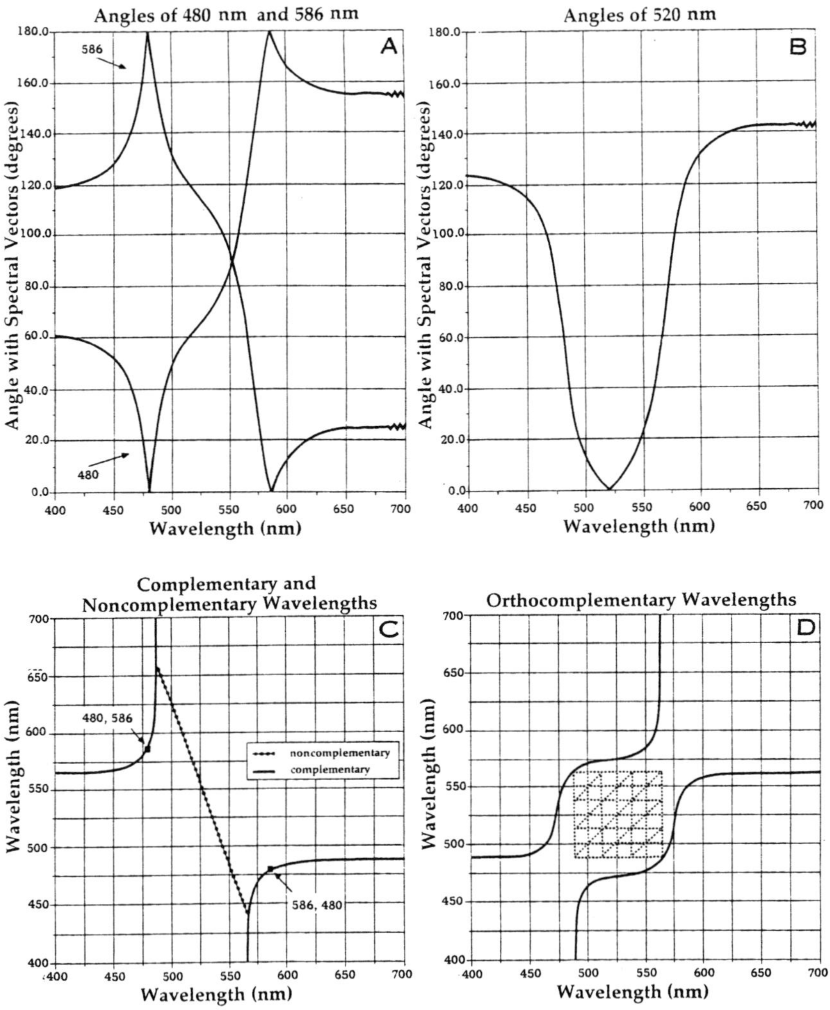

Figure 27. Plots of spectral stimuli, complementary wavelengths, and orthocomplementary wavelengths.

and all other monochromatic vector fundamentals. Since the projected vectors are origin bound, complementary vectors are at 180 degrees with each other. Thus, 480 nm and 586 nm are mutually complementary, potentially mixing to evoke a white sensation. Notice that the functions cross at precisely 90 degrees. *B* plots the angle between the projection of the vector fundamental for monochromatic 520 nm and all other monochromatic fundamentals. The function does not have an angle of 180 degrees with any monochromatic vector fundamental, and therefore, a monochromatic stimulus at 520 nm does not have a potentially complementary stimulus.

I state a generalization: Spectral color stimuli with a spectral complementary have a single spectral orthocomplementary, and spectral color stimuli lacking a spectral complementary have two spectral orthocomplementaries.

Figure 27C, illustrates the locus of spectral complementary wavelengths; the functions are similar to those found in the literature. It also denotes monochromatic stimuli in the central portion of the spectrum lacking monochromatic primaries; these, of course, are the greens, with heterochromatic purples as complementaries.

Figure 27D plots similar relationships between orthocomplementary stimuli—those that are at 90 degrees with each other, in contrast to potentially monochromatic color stimuli that are at 180 degrees with each other. Monochromatic fundamentals with potential complementaries have one orthocomplementary; monochromatic stimuli without potential complementaries have two orthocomplementaries. In *D*, for example, a monochromatic stimulus at 450 nm has a potential monochromatic complementary (see *C*) and one orthocomplementary, that is, 450 nm passes through only one locus in *D*. However, a monochromatic stimulus at 550 nm has no potential monochromatic complementary (see *C*) but two orthocomplementaries, that is, it passes through both loci in *D*.

The Concept of Torque in Complementarism

Figure 28 depicts two complementary spectral color vectors in fundamental color space: 449 nm and 567 nm. The vectors lie on the same plane as the EE white axis, and the vectors potentially compound to evoke a white sensation. As the vectors appear in the diagram, their resultant is not coincident with the equal-energy white vector. To make the resultant coincident, we must either shorten one vector or lengthen the other. How shall we make this adjustment?

We conceive of the two vectors exerting forces as a couple that rotates about the origin, a specific example of a general physical concept known as *torque* or *moments*. When the vectors are in static equilibrium, the torque or moments are equal. Specifically with reference to fundamental color space, the torque is defined as the vector length multiplied by the cosine of the angle between the vector and the chromatic plane. As the vectors are depicted initially in Figure 28, the vectors are indeed in static equilibrium, but the resultant is not the desired white resultant.

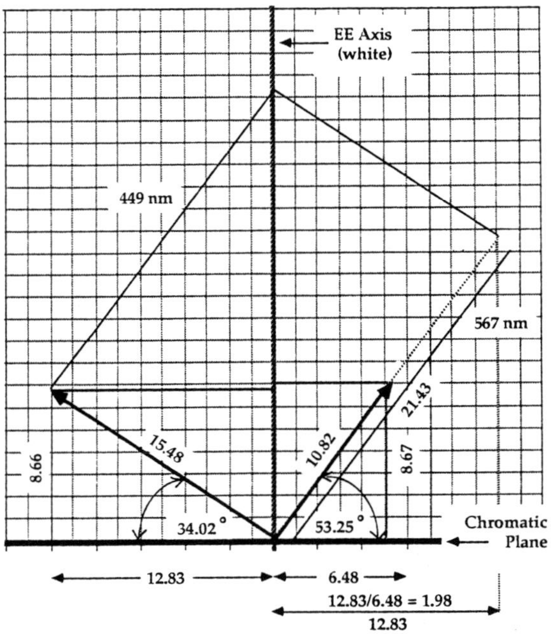

Figure 28. Two complementary spectral color vectors in fundamental color space.

In the analysis of the diagram in Figure 28, all values have been multiplied by 10. The fundamental vectors for 449 nm and 567 nm are potentially complementary because both lie on the plane defined by the two vectors and the equal-energy axis; their vector projections onto the chromatic plane make an angle of 180 degrees. As shown, the fundamental vectors (of these equal-energy stimuli) are of different lengths and also make different angles with the chromatic plane. When added vectorially, the resultant will not be coincident with the equal-energy axis. Therefore, one vector must be adjusted, lengthened or shortened, so that this coincidence occurs. The theory of mechanics states that this will occur when the torques of the potentially complementary vectors are equalized. The vector length of 449 nm is 15.48, the angle between the vector and the chromatic plane is 34.02 degrees, and the torque is represented by the torque vector of length 12.83, coincident with the chromatic plane. Similarly, the vector length of 567 nm is 10.82, the angle between the vector and the chromatic plane is 53.25 degrees, and the torque is represented by the torque vector of length 6.48, coincident with the chromatic plane. We must equalize the respective torques so that the resultant is coincident with the EE white axis.

The ratio of the 449 nm torque to the 567 nm torque is 1.98. Therefore, we extend the length of the 567 nm vector by a factor of 1.98; this is shown in Figure 28. Then, the torque of 449 nm and the extended 567 nm vector are equal, and the two vectors are in static equilibrium with their resultant identical to the EE white axis. Note that the multiplying factor (lengthening the 567 nm vector) is dimensionless and acts precisely as a tristimulus value in a color-matching equation.

Thus, we may assign a torque coefficient—the vector length multiplied by the cosine of the angle with the chromatic plane—to every spectral color stimulus. Figure 29A shows the vector lengths of monochromatic spectral color stimuli on the envelope of the configuration; these are the roots of the diagonals in matrix **R** and were shown in Figure 7A. Every monochromatic vector lies above the chromatic plane and makes an angle with that plane. Figure 29B graphs the cosines of the angle between each monochromatic vector and the chromatic plane. Figure 29C depicts the torque coefficients for the monochromatic vectors, computed by multiplying the function in Figure 29A by the function in Figure 29B—point for point. Note that we may assign torque coefficients even to spectral stimuli without spectral complementaries. The three modes in Figure 29C are extraordinarily well behaved, illustrating a clear mathematical relationship between the monochromatic vector length and their angles with the chromatic plane. The maxima of the torque coefficients have three maxima virtually identical to the densities of the monochromatic vectors projected onto the chromatic plane (compare to Figure 26B). This is not a mathematical property, but rather a characteristic of the color processing mechanism as reflected in the fundamental color space. The modes are somewhat different from those for the vector lengths themselves, shown in Figure 29A. This is especially true of the third peak, now at approximately 605 nm. Here is confirmation of Thornton's (1972) prime colors and Kuehni's (1979) revised division of the spectrum into four segments, where opposing segmental pairs are mutually complementary.

We have effectively decomposed each spectral color vector into two component vectors, the first coincident with the EE white vector and the second on the chromatic plane—the torque, because that is the component that seeks to move the resultant away from the EE white axis. Note that the equal-energy axis and all coincident vectors have zero torque coefficients.

We can now compute the ratios between the torque coefficients of spectral complementary color stimuli, as shown in Figure 29D. These indicate the "power" of each spectral stimulus in neutralizing its complement. As the ratio increases, so does the adjusted length of the vector complement. This may have theoretical importance.

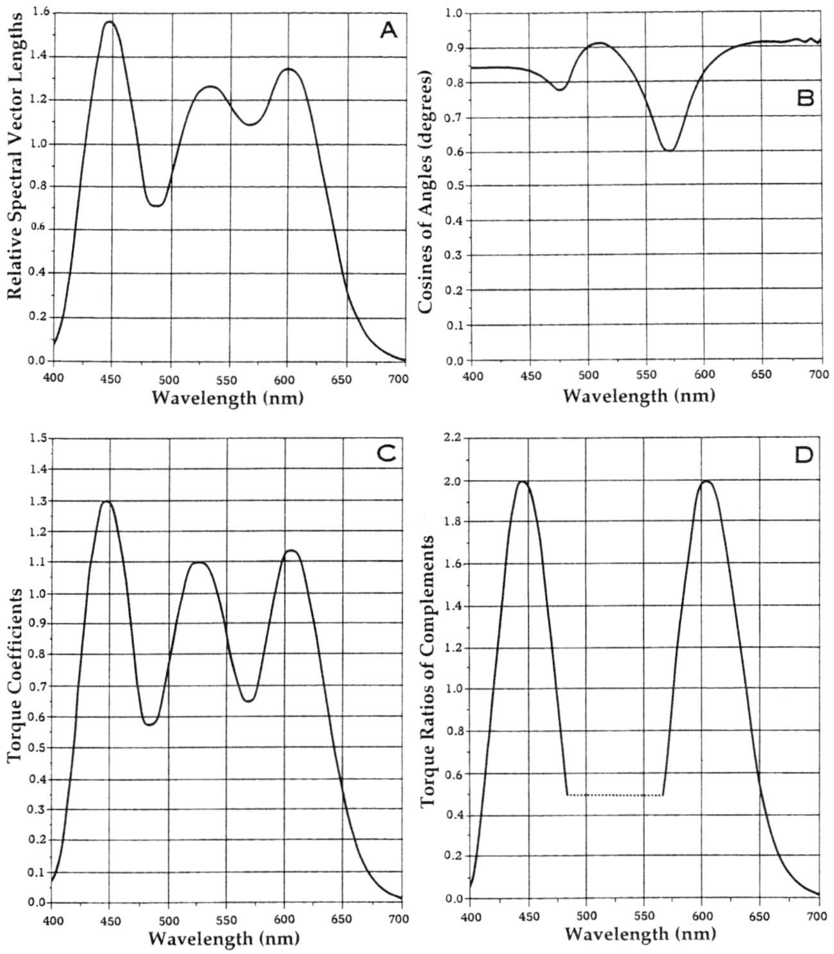

Figure 29. Introduction of the torque concept to fundamental color space and to colorimetry. *Continued next page.*

I was at first surprised at the symmetry of the torque ratios, indicating that both sides of the visible spectrum have equal "rotating" power. This must be true, or the configuration would not be in static equilibrium about the equal-energy vector at the centroid. MacAdam (1938), using the concept of moments (and also the concept of "brightness") with the CIE 1931 chromaticity diagram, found a most asymmetric configuration. He wrote:

> By placing the data on an energy basis, the engineering implications of the curve are more clearly shown. It is seen that, aside from the wave-

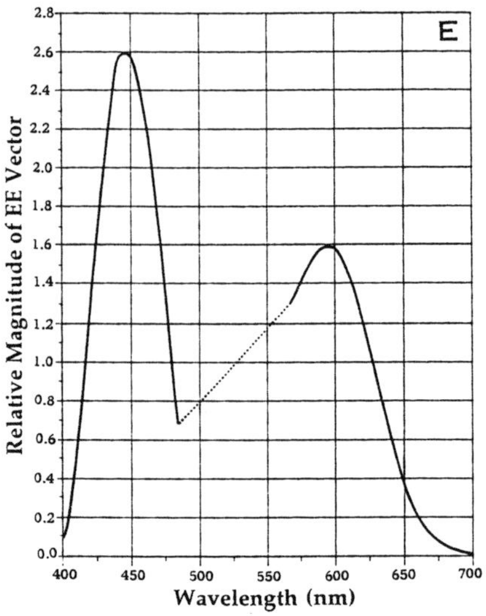

Figure 29. Continued.

> lengths . . . with either extreme red or extreme violet complementaries or no spectrum complementaries whatever, the orange at 605 nm is the most efficient spectrum color in mixtures with its complementary. This orange spectral color is almost twice as effective in neutralizing one watt of its blue-green spectral complementary (490 nm) as is the most effective blue (448nm) in neutralizing one watt of its yellow-green complementary (596 mm).

Although the torque ratio construct correctly describes the lengthening (or shortening) of one member of the complementary pair, it only predicts when the resultant of the pair will be coincident with the EE axis. The magnitude of the resultant, with respect to the EE axis, is generally different for every pair of complementary monochromatic wavelengths. These relative magnitudes are shown in Figure 29E, which plots the magnitude of the equal-energy resultant when the first complementary (designated on the *x*-axis) is held constant and the second complementary is adjusted.

Figure 30 is a dramatic demonstration of the torque coefficients with respect to scalar fundamentals. Figure 30A shows the scalar fundamentals of two potentially complementary spectral stimuli: 480 nm and 586 nm; these are the appropriate rows (or columns) in matrix **R**. Figure 30B shows the sum of these fundamentals; the sum is not the fundamental of a two-peaked equal-energy

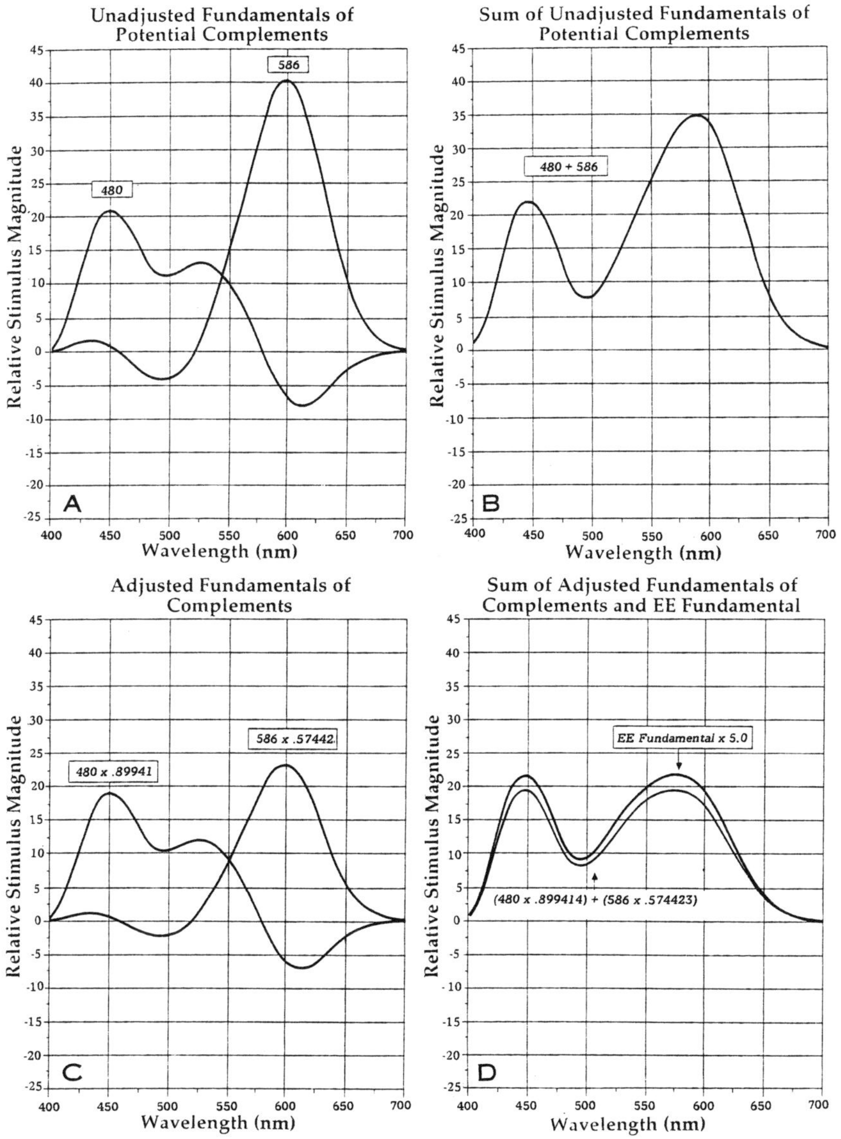

Figure 30. Demonstration of the use of torque coefficients to transform potentially complementary monochromatic color stimuli to absolute complementary monochromatic color stimuli.

fundamental. In *C*, each of the scalar fundamentals in *A* has been weighted by the torque coefficient of the other fundamental; thus, the fundamental for 480 nm has been weighted by .89941 (the torque coefficient of 586 nm) and the fundamental for 580 nm has been weighted by .57442 (the torque coefficient of 480 nm). This "cross-multiplication" procedure guarantees equality of torque for each complementary. The sum of these adjusted complementaries, shown in *D*, adds precisely to the familiar two-peaked fundamental of an equal-energy fundamental. For comparison, *D* also illustrates the fundamental of an equal-energy fundamental at a slightly greater magnitude.

14

The Near-Orthogonal Fundamentals of Complementarism

The visible spectrum can be divided into three segments based on the phenomenon of complementarism. The first segment (the blues) ranges from 401 nm to 487 nm, the second segment (the greens) ranges from 488 nm to 564 nm, and the third segment (the reds) ranges from 565 nm to 700 nm. Every monochromatic stimulus (wavelength) in the first segment has a complementary stimulus (wavelength) in the third segment, and every monochromatic stimulus (wavelength) in the third segment has a complementary monochromatic stimulus (wavelength) in the first segment. Each monochromatic stimulus in the first and third segments is associated with a single orthocomplementary monochromatic stimulus. Monochromatic stimuli (wavelengths) in the second segment have no spectral complementaries, although they do have compounded purple complementaries. However, each monochromatic stimulus in the second segment has two orthocomplementary spectral stimuli, which are complementary to each other.

This tripartite division of the spectrum suggests that the three segments might serve as heterochromatic primaries, a basis for fundamental color space. Accordingly, I computed the coordinates (in the canonical orthonormal frame in fundamental color space) of the combined monochromatic stimuli in each segment. I then transformed the configuration to reference these primaries of complementarism. A computer rendition appears in Figure 31. The sums of the fundamentals in each segment are heterochromatic fundamentals that are closely

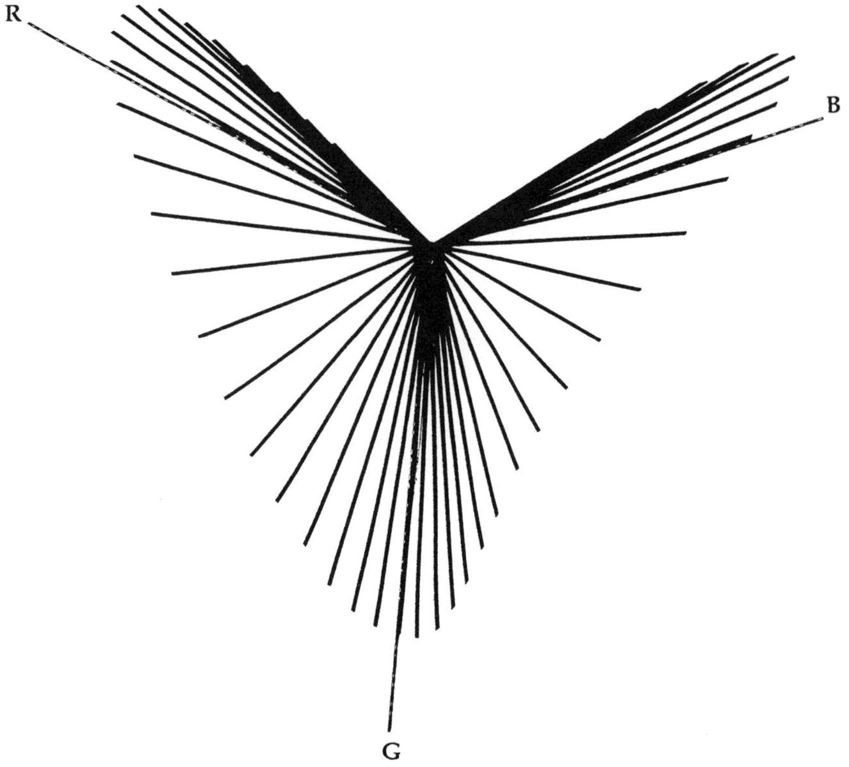

Figure 31. An interesting partition of an equal-energy spectrum using the monochromatic projections onto the chromatic plane.

orthogonal and normalized to form a near-orthonormal matrix. In this view, the positioning of these reference vectors (axes) is viewed from the equal-energy axis at the centroid of the configuration. The axes are within the configuration but rather close to the spectral envelope.

These reference vectors, based on the natural division of the spectrum into its three parts, are closely orthonormal. Taken together, they form a nonnormalized matrix $\mathbf{F}_c$. When $\mathbf{F}_c$ is normalized, $\mathbf{F}_c'\mathbf{F}_c$ should approach an identity matrix; here is that matrix comprising direction cosines:

$$\begin{bmatrix} 1.0000 & -.0602 & -.0008 \\ -.0602 & 1.0000 & .0059 \\ -.0008 & .0059 & 1.0000 \end{bmatrix}$$

Thus, the angles between the reference axes of complementarism are 93.45 degrees, 90.05 degrees, and 89.66 degrees.

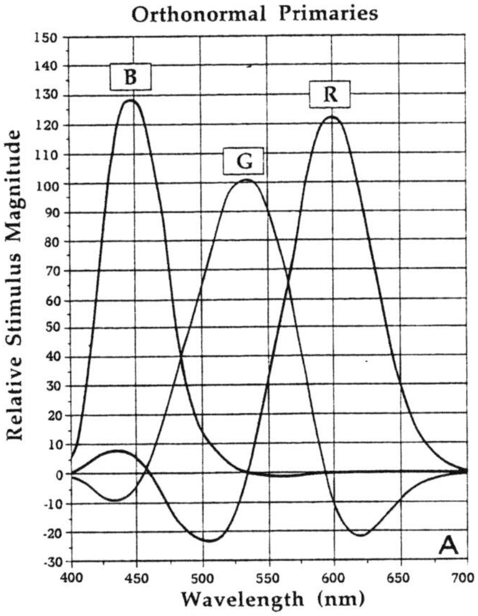

Figure 32. Interesting properties of a particular orthonormal matrix. *Continued next page.*

The scalar fundamentals of these closely orthonormal functions are shown in Figure 32. This interesting triplet (Figure 32A) may have consequences for printing and the graphic arts, and also for high-definition television based on a subtractive color mixture. Any two primaries are complementary to the third. The sum of the red and blue primaries is shown; this is the printer's magenta. The sum of the green and red primaries is shown; this is the printer's yellow. The sum of the blue and the green primaries is shown; this is the printer's cyan. The sum of the blue, green, and red primaries is precisely the fundamental of a white equal-energy source.

I now speak of subtractive colorimetry, derived from additive colorimetric data. Subtracting the magenta from the white gives green. Subtracting the yellow from the white gives blue. Subtracting the cyan from the white gives red. Black prevails when all three are subtracted from white.

The tripartite division of the spectrum has intrigued reseachers since the earliest days of photography (see, for example, Hübl, 1897). My proposed solution to tripartite division, based on spectral complementaries, is closely related to another tripartite division based on the trace of matrix **R**, previously reported by Cohen and Kappauf (1982). The trace, the sum of the diagonals in any matrix, is precisely equal to the dimensionality of the idempotent matrix. In the

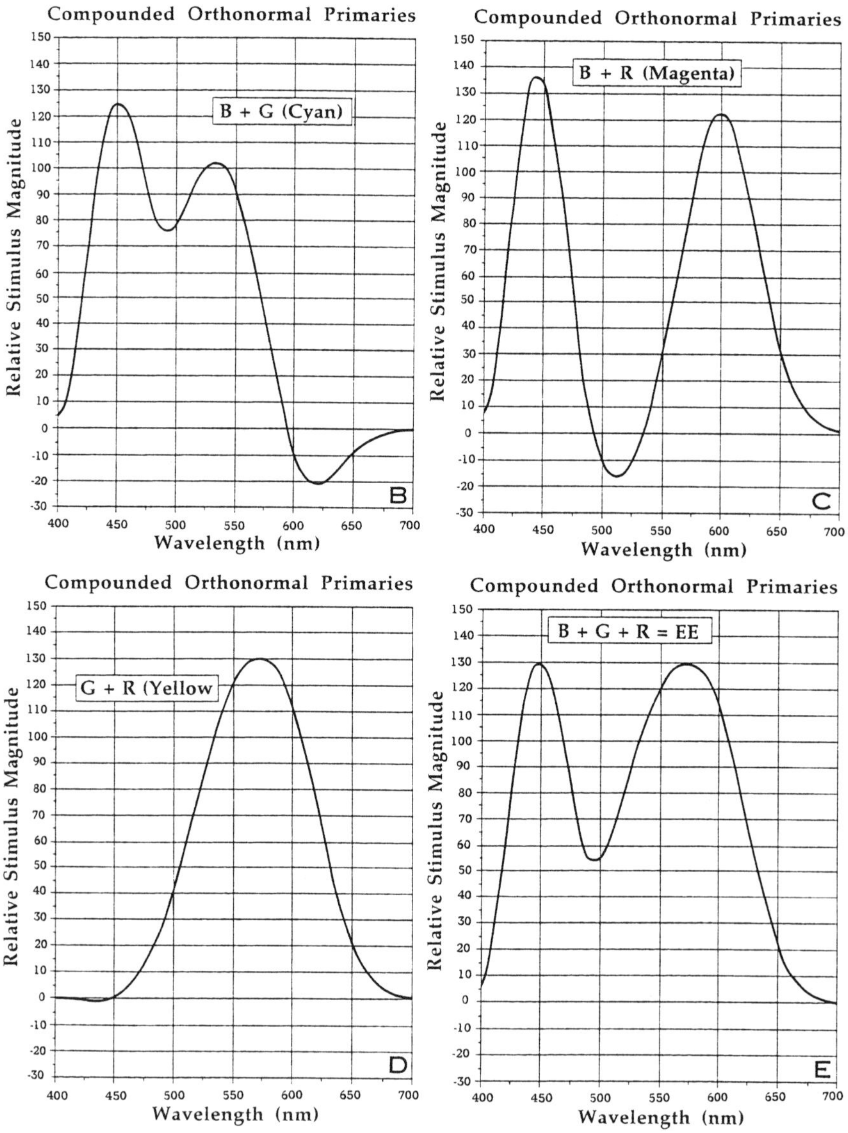

Figure 32. Continued.

case of matrix **R**, the trace is precisely equal to three. The trace is a measure of variance, and the cumulative sum of diagonals in matrix **R** is shown in Figure 7C. The trace divides the spectrum into three parts, at summation point 1.00 and summation point 2.00, each segment with equal variance, termed *trace segments.*

The first portion is delimited by wavelengths 401 nm to 480 nm, the second is delimited by wavelengths 481 nm to 564 nm, and the third is delimited by wavelengths 565 nm to 700 nm. These segments are close to those emerging in the study of stimuli with and without complementaries. I have summed these three segments respectively, and three fundamentals have emerged. These fundamentals are almost identical to the fundamentals of complementarism in Figure 32 and also near-orthogonal.

The following matrix contains the variance contributed by each column in matrix **F**, with respect to the three trace segments:

	I	II	III
Variance from 401 to 480	.977937	.022566	.001294
Variance from 481 to 564	.021575	.858252	.123397
Variance from 565 to 700	.000488	.119142	.875309

This matrix is virtually symmetrical. Each row and each column adds to unity, as a mathematical necessity. This is the matrix multiplied by its transpose:

$$\begin{bmatrix} 1.00000 & .00000 & .00000 \\ .00000 & 1.00000 & .00000 \\ .00000 & .00000 & 1.00000 \end{bmatrix}$$

Thus, the variance matrix is an orthogonal matrix because its transpose equals its inverse. This property may have application in the electronic transmission of colorimetric information.

Color Specification with Matrix-R Operations

Here I disclose three separate strategies for the specification of any color stimulus, real or imaginary, and its evoked color sensation, real or imaginary: (1) *tristimulus values* 3×1 matrix **Q**, (2) *tricolor values* 3×1 matrix **V**, and (3) *tricolor coordinates* 3×1 matrix **S**. Tristimulus values, of course, were popularized by the CIE and are well known. Cohen and Kappauf (1982) introduced tricolor values and tricolor coordinates.

These three types of specifications are all represented unequivocally by simple, basic equations of matrix-**R** operations.

Types of Specification

Tristimulus Values

These specifiers, entries in 3×1 matrix **Q**, are defined as

$$\mathbf{Q} = \mathbf{A}'\mathbf{N} = \mathbf{A}'\mathbf{N}^*. \tag{14.1}$$

Tristimulus values, referring to three arbitrary primaries, theoretically result directly from a color-matching experiment and are the row entries in an **A** matrix or a linear combination thereof, as equation 14.1 implies.

In the valid color-matching equation, 6.4, tristimulus values are weights operating on the fundamentals (or radiometric functions) of the three primaries—say, a primary named X, a primary named Y, and a primary named Z. The CIE X,Y,Z tristimulus values, for example, have become a de facto standard, having been officially recommended by the CIE long before the projection operator and linear algebra became integral components in the curricula of higher mathematics. As demonstrated repeatedly, tristimulus values are not the "quantities" of three primary stimuli required to effect a metameric match, as propagated by the CIE. Equal tristimulus values do not mean that equal magnitudes of the primaries X,Y,Z are required to match the sensation evoked by an equal-energy stimulus; in fact, equal magnitudes are not required—despite the apparent "equality" of the respective tristimulus values. The magnitudes (in irradiance units) of X, Y, and Z are far from equal, as already demonstrated.

Tristimulus values are merely dimensionless numbers weighting the fundamentals of three primaries. The fundamentals (not the tristimulus values) may be represented as scalars or vectors. (See Figures 34–36.) When so weighted, the sum of the fundamentals representing the primaries is the fundamental of the color stimulus matched. The fundamental before weighting is designated the *initial* fundamental; the fundamental after weighting is designated the *terminal* fundamental. Depending on the arbitrary choice of the initial fundamentals representing the primaries, the tristimulus values may be any number whatever; tristimulus values have no meaning until they multiply the initial primary fundamentals (scalar or vector) and the terminal primary fundamentals (scalar or vector) emerge. The terminal magnitudes of the primary fundamentals are inviolate in the color-matching experiment, in contrast to the tristimulus values themselves.

Because the tristimulus values are merely multipliers or weights and because they are solely at the discretion of the investigator, they may be summed only when they weight the same initial fundamental, say, the fundamental of the primary named X. Tristimulus values cannot be logically added when they weight different initial primaries, say those named X, Y, and Z, respectively, although this is a current pervasive divertissement. For a pictorial demonstration of this continuing futilty, see Cohen (1988).

Tristimulus values do not appear as entities in the fundamental color space evolved in this book.

Tricolor Values

These specifiers are entries in 3 × 1 matrix **V** and are defined

$$\mathbf{V} = \mathbf{E'N} = \mathbf{E'N^*}. \tag{14.2}$$

Matrix **E** is derived from matrix **A** by the now-familiar equation $\mathbf{E} = \mathbf{A}(\mathbf{A'A})^{-1}$. Tricolor values are the row entries in matrix **E**, or a linear combination thereof, as implied by equation 14.2.

Tricolor values specify any given color fundamental (and the evoked sensation) represented by a terminal vector in fundamental color space. Tricolor values are three scalar products between the given vector and usually (although not necessarily) three arbitrary vectors selected from the continuum of spectral vectors lying on the envelope of fundamental color space. Stated differently, tristimulus values of a given color sensation are three arbitrary points chosen from its fundamental, the arbitrary points representing preselected wavelengths in the visible spectrum.

In principle, the three reference vectors need not be spectral vectors, but they may be any independent vectors—as those within the configuration representing real color fundamentals and their evoked real color sensations, or those outside the configuration represent imaginary color fundamentals and their evoked imaginary color sensations. However, reference to real spectral stimuli has unique advantages.

Tricolor Coordinates

These specifiers are entries in 3 × 1 matrix **S** and are defined

$$\mathbf{S} = \mathbf{F'N} = \mathbf{F'N^*}, \tag{14.3}$$

where matrix **F** is an orthonormal matrix, defined by the now-familiar equations $\mathbf{A}(\mathbf{A'A})^{-1}\mathbf{A'} = \mathbf{AGG'A'} = \mathbf{R}$, and $\mathbf{AG} = \mathbf{F}$, and $\mathbf{FF'} = \mathbf{R}$. Tricolor values are the row entries in matrix **F**, or a linear combination thereof, as implied by equation 14.3.

Tricolor coordinates specify any given color fundamental (and its evoked color sensation) represented by a terminal vector in fundamental color space.

Tricolor values are three orthogonal projections of the given vector onto three arbitrary orthogonal reference axes (vectors). The arbitrary reference frame is obtained by one of the procedures already discussed—by Gram-Schmidt or by matrix-**R** operations themselves. Usually the first axis is passed through an important vector, say, that representing an equal-energy stimulus, and the remaining two axes are orthogonal to each other and lie on a plane orthogonal to the first. I emphasize, once again, that an orthogonal frame is not under the control of the experimenter; in fundamental color space, an investigator cannot transform three oblique vectors into three orthogonal vectors.

Occasionally three "natural" orthogonal vectors emerge from the fundamental vector configuration, as that triplet comprising three spectral vectors at 455 nm, 513 nm, and 584 nm. This and similar triplets may be the orthogonal vectors used as primaries in an orthonormal **F** matrix.

15

The Mathematical Coalition of Color Specification

Here is the now-familiar equation of the projection operator (matrix **R**) developed earlier:

$$\mathbf{A}(\mathbf{A}'\mathbf{A})^{-1}\mathbf{A}'\mathbf{N} = \mathbf{N}^* = \mathbf{RN}. \tag{15.1}$$

This equation is the mathematics of the Wyszecki hypothesis. Note that it is written without internal brackets.

In my development, this equation has occurred repeatedly and is arguably the most potent equation in contemporary colorimetry. This identical equation specifies all three color specifications of color stimuli scalar and vector fundamentals and their evoked color sensations. My demonstration will use either scalar or vector fundamentals, with primary consideration given to clarity. All specifications employ this same equation, differing only by placement of brackets.

Here is the critical equation is written with the first placement of brackets:

$$[\mathbf{A}(\mathbf{A}'\mathbf{A})^{-1}][\mathbf{A}'\mathbf{N}] = \mathbf{N}^*. \tag{15.2}$$

But $\mathbf{Q} = [\mathbf{A}'\mathbf{N}]$ and $\mathbf{E} = [\mathbf{A}(\mathbf{A}'\mathbf{A})^{-1}]$; therefore, premultiplication of matrix **Q** (tristimulus values) by matrix **E** (fundamentals of the primaries) brings recovery of the $\mathbf{N}^*$ fundamental. This is, of course, a statement of the valid color-matching equation, 6.4. Here is the critical equation written with the second placement of brackets:

$$[\mathbf{A}][(\mathbf{A}'\mathbf{A})^{-1}\mathbf{A}'\mathbf{N}] = \mathbf{N}^*. \tag{15.3}$$

But $\mathbf{V} = \mathbf{E}'\mathbf{N} = [(\mathbf{A}'\mathbf{A})^{-1}\mathbf{A}'\mathbf{N}]$; therefore, premultiplication of matrix **V** (tricolor values) by matrix **A** (color-matching functions emerging in a color-matching experiment) brings recovery of the $\mathbf{N}^*$ fundamental.

The same critical equation is written with the third placement of brackets:

$$[\mathbf{F}][\mathbf{F}'\mathbf{N}] = \mathbf{N}^*. \tag{15.4}$$

Cohen and Kappauf (1982) reported that the value $(\mathbf{A}'\mathbf{A})^{-1}$, labeled $\mathbf{M}^{-1}{}_{e}$, is gramian and may be resolved into $\mathbf{GG}'$. Therefore, equation 15.4 may be written $\mathbf{A}(\mathbf{A}'\mathbf{A})^{-1}\mathbf{N} = \mathbf{AGG}'\mathbf{AN}' = \mathbf{FF}'\mathbf{N} = \mathbf{RN} = \mathbf{N}^*$, where **F** is an orthonormal matrix. But $\mathbf{SS} = \mathbf{F}'\mathbf{N}$; therefore, premultiplication of matrix **SS**, tricolor coordinates, by matrix **F** (orthonormal color functions) brings recovery of the fundamental $\mathbf{N}^*$.

Three color specification devices—tristimulus values, tricolor values, and tricolor coordinates—may be transformed one to another. One matrix **Q** can be transformed to another matrix $\mathbf{Q}_1$ with different primaries, and one matrix **SS** may be transformed to another matrix $\mathbf{A}_1$ with different primaries. One matrix **Q** may be transformed to a matrix **SS** ($\mathbf{M}_e\mathbf{Q} = \mathbf{SS}$), and similarly, one matrix **SS** may be transformed to a matrix **Q** ($\mathbf{M}_a\mathbf{SS} = \mathbf{Q}$). However, transformation to orthogonal tricolor coordinates is possible only with inherently orthogonal primaries, although specification in one orthogonal frame **V** may be transformed to specification in another orthogonal frame $\mathbf{V}_1$.

Some color specification devices have advantages over others. Only tristimulus values are currently in use, but of all specifiers, these are probably weakest because valid mathematical operations are extremely limited. Tristimulus values are merely dimensionless weights placed on the fundamental of the primaries. Tristimulus values have a vector representation in so-called affine space, where distance and angle are meaningless, in contrast to the Euclidean nature of fundamental color space.

Tricolor values are scalar products, three arbitrary values from the fundamental—which is itself a series of scalar products. Every color stimulus (and its unique sensation) is represented by its scalar fundamental as originally conceived by Wyszecki, and its corresponding vector fundamental metamer is in the immutable fundamental color space. Tricolor values are the scalar products between the vector fundamental of the color to be specified and the vector fundamentals of three arbitrary chosen spectral vectors. This guarantees that tricolor values are values taken from the scalar fundamentals. This triplet of tricolor values gives some indication of the entire scalar fundamental metamer. (Note that the scalar products may be between the vector fundamental metamer for the color to be specified and any three nonspectral independent vectors within the space; these scalar products will not be values taken from the scalar fundamental but may prove advantageous.) Scalar products are additive; one may be added to another.

Tricolor values have a total independence. If one primary is changed, only the tricolor value for that primary changes. The tricolor values for the remaining two primaries remain constant, although all three primaries are highly correlated. Contrast with the dramatic changes reflected in tristimulus values under these same conditions.

Tricolor coordinates are described by Euclidean geometry in general and Cartesian analytical geometry in particular (including polar, spherical, and cylindrical coordinates). Orthogonal tricolor coordinates are independent. All theorems of analytical geometry may be applied with ease, and all are valid.

The CIE and Fundamental Color Space

The CIE has provided de facto standards for color-matching data—the 1931 standard observer based on a 2 degree field and the 1964 standard observer based on a 10 degree field. This book uses the 1964 data almost exclusively.

The CIE began its adoptions not with colorimetric data but rather with a standard for luminous efficiency within an investigative area termed *photometry*. Clearly some radiations evoke visual sensations, and other radiations do not. Furthermore, some radiations of equal power evoke "brighter" sensations than other radiations. The CIE began with standards for photometry, and these same standards then became part of its 1931 standard observer. The two domains—the photometry of color-sensation brightness and the model for color-matching—were combined and have remained inextricably integrated.

Luminance and Brightness (Lightness)

The study of photometry has two parts: light dispersion and illumination, and color sensation magnitude as a function of stimulus magnitude.

The first part, termed *physical photometry* and concerned with light dispersion and illumination, was addressed by two classical works: Pierre Bouguer (1760) and Johann Heinrich Lambert (1760). (A remembrance of Lambert's work survives; a *lambert* is a unit of luminance.) These volumes described now commonly known general photometric laws for light distribution: as the numerical addition of illumination, the inverse square law, and the cosine law of emission. Contemporary physical photometry is described in several modern volumes, including Walsh (1963), Tarrant (1982), and the CIE (1983). I am not concerned with this aspect of photometry in this book.

However, *heterochromatic photometry*—the measurement of color sensation magnitudes evoked by color stimuli of varying wavelengths and intensities—is of critical importance. Light comprises the visible spectrum by definition and ranges from about 380 nm to 770 nm specified in wavelength where the unit is a nanometer; light is sometimes specified in frequency, similar to sound radiation, and the suggested unit is a *fresnel*, a correlate of the reciprocal of wavelength, whose units are 10^{-12} cycles per second. This comes as a recommendation by Wald (1965). Today in the face of a well-defined metric system of measure, we would call these units *terahertz*.

Light is but one type of radiation within the vast electromagnetic spectrum, with wavelengths from kilometers to ten-millionths of a meter. No detector is sensitive to all radiations at all wavelengths but many instruments, based on different principles, are sensitive to portions of the range. Some detectors are resonators, some are radiometers, and some depend on the sensitivities of photographic emulsions. One uniquely important detector is the color processing mechanism itself. Thermal detectors have the widest limits and the color processing mechanism the narrowest limits—less than a single octave.

Magnitude of Color Sensations

The quantity or *magnitude* of a color sensation has no easily communicable meaning; I can imagine nothing more subjective than a color sensation. At the inception of heterochromatic photometry, *brightness* was defined intuitively as a measure of sensation magnitude, and *luminosity* was simply another word for *brightness*. Much later, *brightness* was restricted to the magnitude of transmittance stimuli, and *lightness* was restricted to the magnitude of reflectance stimuli; here I use them interchangeably.

Some visible radiations (wavelengths) are "better" than other radiations because some evoke color sensations of greater "magnitude." These "better" radiations are presumed to be described by luminosity coefficients of greater magnitude and therefore are said to be more efficient. Although Isaac Newton was unaware of the breadth of the electromagnetic spectrum, he did observe that some spectral color stimuli evoked brighter color sensations than others. In the first edition of his *Opticks*, Newton (1704) discussed techniques to minimize the effects of chromatic lens aberration. He wrote: "But it is further to be noted, that the most luminous [*sic*] of the prismatick Colours are the Yellow and Orange. These affect the Senses more strongly than all the rest together, and next to these in strength are the Red and Green. The Blue compared with these is a faint and dark Colour, and the Indigo and Violet are much darker and fainter, so that these compared with the stronger Colours are little to be regarded." So began the discipline of heterochromatic photometry.

Newton's observations were confirmed by Joseph von Fraunhofer, who discovered the absorption lines on the solar spectrum bearing his name. In 1815, Fraunhofer was confronted with Newton's earlier problem: the correction of chromatic aberration in lenses. Essentially Fraunhofer (1814–1815) used a Maxwellian field with a spectral stimulus on one side and a white stimulus on the other. He systematically adjusted the white stimulus until the corresponding white sensation was matched in brightness to each of the corresponding colored spectral sensations. Fraunhofer's results approached current luminosity functions, with maximum sensitivity toward the center of the visible spectrum. The results were confirmed essentially by Vierordt (1869) and Dobrowolsky (1872). These experiments, however, are merely suggestive because these investigators had no means of measuring the absolute energy of their comparison spectral color stimuli; implicitly, they had assumed that the solar spectrum was an equal-energy spectrum, which, of course, is untrue.

The Purkinje Phenomenon

In 1825, the problem of heterochromatic photometry became more complex with the discovery of the well-known Purkinje phenomenon. The original publication by Jan Evangelista Purkinje (1825) was confined to a disappointing subjective description in three paragraphs. Purkinje reported that as night turns to day, different color sensations emerge differentially. At very low illumination levels, all chromatic stimuli evoke only achromatic sensations, ranging from near-black to near-white; red evokes the darkest sensation. As illumination is increased, blue stimuli evoke blue sensations, and as illumination is further increased, other spectral stimuli evoke their hues; finally, red stimuli evoke red sensations. Stated obversely, as the irradiance level is reduced, the maximal relative brightness of spectral sensations shifts from green toward blue. The Purkinje shift was confirmed by Seebeck (1837), Dove (1852), and Grailich (1854).

In modern terms, the *photopic luminosity function* refers to the relative efficiency of light at high irradiance levels, and the *scotopic luminosity function* refers to the relative efficiency of light at low irradiance levels. In this work, I am concerned with the photopic luminosity function almost exclusively.

Given the differential sensitivity of light as a function of wavelength and illumination level (Purkinje shift), the definition of light is extremely complex.

Luminosity Functions

For a given level of illumination, heterochromatic photometry provides that a luminosity coefficient, designated $V_{(k)}$, is associated with each monochromatic color stimulus (at unit irradiance) spanning the visible spectrum. (Luminosity coefficients were formerly called visibility coefficients—hence the V.) The continuum of these luminosity coefficients comprises an uninterrupted, single-valued function with its maximum toward the center of the visual spectrum and its minimum at the limits.

Luminosity coefficients were presumed measures of subjective brightness. The efficiency (brightness or luminance per watt) of light $V_{(k)}$ is the ratio of luminosity (psychological brightness or lightness) to radiant power (energy) $E_{(k)}$ at wavelengths spanning the spectrum. Thus, we have $V_{(k)} = L_{(k)}/E_{(k)}$. The reciprocal of $V_{(k)}$ is often plotted, and sometimes its logarithm. Luminous efficiency is usually expressed in relative units (maximum = 1.00) but may be expressed in absolute units.

Measurement of Luminosity

The luminosity function is gauged with extraordinary difficulty with five general techniques:

1. *Equality of brightness.* The observer equates the subjective brightness of one sensation (evoked by a given stimulus at one wavelength) to the subjective brightness of a second sensation (evoked by another stimulus at another wavelength). When the wavelengths of the stimuli are adjacent or almost adjacent, the method is termed *small-step* or *cascade*, and the observer has minimal difficulty so long as the wavelength differences are below or almost below threshold. When the stimuli are not closely contiguous with respect to wavelength, the method is termed *large-step*, and the observer exhibits extreme variability between successive matches—and the judgments of different observers varies considerably. Helmholtz denied the validity of the large-step technique. This method has been used, for example, by Ives and described in his 1912 series of papers.

2. *Visual acuity.* The observer sequentially adjusts the physical irradiance of monochromatic stimuli until the monochromatic illumination permits visual resolution of fine detail as, for example, the small numbers in a logarithm table. This method has been used, for example, by Langley (1889).

3. *Critical frequency.* This technique depends on the visual phenomenon termed *persistence of vision.* Monochromatic lights are viewed sequentially through a rotating shutter; that is, the monochromatic stimulus is alternated

with a black stimulus. The observer adjusts the speed of the rotating shutter until flicker disappears. Presumably brighter color sensations have longer visual persistence. When critical shutter speeds are equal for two monochromatic stimuli, equal luminance is attributed to the stimuli. More commonly, when the critical shutter speeds are unequal, theory predicts the respective luminance values. This method has been used, for example, by Ives and is discussed in his 1912 papers.

4. *Flicker photometer.* Two different monochromatic stimuli are alternated on the same field. The observer adjusts the irradiance of one or the other stimuli until flicker disappears. At that point, equal luminances are attributed to the stimuli. This method has been used, for example, by Nutting (1908–1909).

5. *Threshold of vision.* The limen or threshold of a stimulus is assumed to be highly correlated with its relative luminance. The observer views a subthreshold stimulus evoking a black sensation. Then the observer increases the stimulus irradiance until the stimulus is at threshold—that is, evokes a minimal color sensation. The usual psychophysical techniques are involved; sometimes the stimulus is suprathreshold, and the observer reduces the stimulus irradiance to bring the sensation to threshold. The final luminosity function comprises complicated statistical entities, for a limen is invariably a number derived by statistical analysis. This method was used, for example, by Pflüger (1902).

As an example of luminosity measurement in general and the method of visual acuity in particular, I turn to the work of Samuel P. Langley (1889), best remembered for his early research on heavier-than-air flight. The U.S. Air Force's Langley Field remains a remembrance of and tribute to him. Langley, however, possessed other interests, having invented the bolometer, a device for measuring minute amounts of radiant energy by resistance changes caused by small temperature changes.

On April 19, 1888, Langley addressed the National Academy of Sciences in Washington on what then was an obscure topic, energy and vision:

> While it is quite familiar that the luminosity [*sic*] of any spectral ray, and indeed it is but another manifestation of the same energy, I have recently had occasion to notice that there is . . . a failure to recognize how totally different optical effects may be produced by one and the same amount of energy according to the wave-length in which this energy is exhibited. . . . I have undertaken, therefore, an experimental investigation. . . .
>
> We shall evidently need two correlated sets of experiments; the first set to determine the amount of energy in each ray, and second to show the corresponding visual effect.
>
> For the first of these, we must select that manifestation which disguises it least; and in this respect beyond comparison the thermal one

> stands first, as the *heat* dispersed by a glass prism and shown by a thickly lamp-blacked thermometric apparatus [bolometer] is, throughout the visible spectrum, very nearly proportionate to the energy itself. . . .
>
> The object of these experiments, then, is to take some one constant amount of energy, to actually or virtually display it successively in different portions of the spectrum, and to observe in what proportion the optical or visual effects of this fixed amount of energy vary, according to the wavelength in which it is conveyed.

Langley's test was visual acuity under successive wavelengths. His observers read numbers from a table of logarithms, and the number of correct numbers was a measure of luminance. Langley's final results indicate maximum luminance (not necessarily brightness) to be in the green segment of the spectrum, about 530 nm. He wrote: "The visual effect produced by any given amount of energy varies enormously according the light in question. . . . It appears that the same amount of energy may produce at least 100,000 times the visual effect in one color of the spectrum than it does in another."

Numerous peripheral factors confound the measurement of the luminosity function: the Purkinje shift, the size of the stimulus field, the surround of the stimulus field, and so forth. Some of these peripheral factors have been researched—for example, by Hurvich and Jameson (1953a, 1953b, 1954) and Walters and Wright (1943). I am *not* concerned with these factors here.

Development of Luminosity Functions and Adoption by the CIE

The first significant luminosity data were gathered by Arthur König (1891), a distinguished student of the even more distinguished Herman von Helmholtz. König's paper appeared in an 1891 festschrift honoring Helmholtz on his sixtieth birthday.

Thereafter, research was largely confined to laboratories in the United States —particularly the U.S. National Bureau of Standards, National Electric Lamp Association (NELA) Park Laboratories, and the Bell Telephone Laboratories. The motivation was the immediate development and distribution of electric lamps and the eventual electronic transmission of visual information. The prime investigator in this movement was Herbert Eugene Ives (not to be confused with his father, Frederic Eugene Ives), honored by the annual medal awarded by the Optical Society of America. Ives (1912a, 1912b, 1912c, 1912d, 1912e) wrote on the general problem of the photometry of lights of different colors, Ives and Kingsbury (1914, 1916) on the theory of the flicker photometer, and P. G. Nutting (1908–1909, 1911, 1915) on luminous equivalent of radiation. (Nutting recalculated König's data, to obtain somewhat modified results.) Other contributions were made by Hyde and Forsythe (1915), Hyde, Forsythe, and Cady (1918), Crittenden and Richtmeyer (1918–1919), and Coblentz and Emerson (1918–1919).

Leonard T. Troland (1922), a co-inventor of the technicolor process, wrote and published the now-famous *Report of the Committee on Colorimetry for 1920–1921*, commissioned by the Optical Society of America. Troland recorded that a luminosity (visibility) function (at 10 millimicron intervals from 380 nm to 780 nm) had been adopted as standard by the Illuminating Engineering Society and also as standard by the Optical Society of America. In 1924, these same data were adopted as standard by the CIE. Of this, W. D. Wright (1958) has written, "All heterochromatic photometry which made any claim to be standard had to be in close conformity with this curve."

The basis for the adoption has been carefully reviewed by Kasson S. Gibson (1948). The CIE adoption was based essentially on measurements made by Gibson and Tyndall (1923–1924) at the U.S. Bureau of Standards. At the end of their own experiments, these researchers compiled all existing photopic luminosity data derived from experiments using small-step equality-of-brightness and flicker techniques and also based on more than two hundred observers. From these and their own data derived from additional research, Kasson and Gibson derived recommended averages. These data were presented at the Sixth Meeting of the International Commission on Illumination at Geneva in 1924.

When these data were adopted by the CIE in 1924 as the correct luminosity data by definition, the function was effectively cast in concrete. Like an ecclesiastical decree, inquiry was inhibited, and change was tortuous. In fact, the CIE luminosity function has not been revised or altered in nearly 75 years, and the original function still remains the official recommendation. In 1931, these same data were incorporated (as the Y color matching function) into the 1931 standard observer for colorimetry, again adopted officially by the International Commission on Illumination.

The shape of this official photopic luminosity function is well known; its maximum is at 555 nm, and it has a somewhat asymmetrical bell shape—something akin to the normal probability function. The photopic luminosity function falls to zero at the limits of the visible spectrum by definition—about 380 nm and 770 nm. Several researchers have attempted to write analytical representations of the luminosity function: Goldhammer (1905), Nutting (1908–1909), and Kingsbury (1916) on luminosity data before the 1924 adoption, and Tyndall and Gibson (1924), Walsh (1925), Houston (1930, 1931, 1932), and Moon and Spencer (1933) on the CIE standard function.

Analytical approximations, however, do not provide sufficient accuracy for accurate interpolation; computational methods have been employed. Walsh (1925) provided interpolated data, mostly to three significant figures. D. B. Judd (1931) extended the standard visibility function to intervals of 1 millimicron by third-difference osculatory interpolation, to four significant figures. Yamauti and Okamatu (1935) provided similar interpolated data.

E. O. Hulbert (1926) has suggested that the shape of the photopic luminosity function is roughly similar to the transmission characteristics of seawater; that is, the maximum transmission of seawater is at about 555 nm, and it transmits nothing beyond the visible spectrum. Hulbert has surmised that the luminosity function evolved in the primordial ooze of the Paleozoic age, before our form of life developed appendages to move about the relatively dry earth. The luminosity function, according to Hulbert, did not evolve further. This resemblance may be pure coincidence or, in fact, have evolutionary meaning. Note, however, that the luminosity function is a color-matching function, while the radiometric function of seawater is a fundamental; in general, these are different constructs.

Presumed Properties of CIE Luminosity

Psychophysicists have long attended to the scalar additivity of luminances or brightnesses associated with different color stimuli. Ives (1912a) wrote, "A well-nigh indispensable requirement of a method of measurement is that its renderings shall conform to geometrical axioms. The first of these is that things equal to the same thing should be equal to each other; the second, that the whole shall be equal to the sum of its parts." Ives concluded that only measurements made by the flicker method conform to these requirements. Nevertheless, the CIE has embraced these axioms with respect to the adopted function as does classical heterochromatic photometry. (On this subject, see LeGrand, 1957, 1972.)

Given two monochromatic stimuli E_1 and E_2 at equal irradiance and their associated luminosity coefficients V_1 and V_2, the magnitude of the color sensation evoked by the first is given by E_1V_1, and the magnitude of the second is given by E_2V_2. When V_1 is twice the value of V_2, classical color photometry predicts that the first color sensation will be twice as bright as the second. Twice as bright? Given two color sensations with widely different hues, side by side, when does an observer conclude that the first is precisely twice as bright as the second? In over a century, researchers have concluded that this judgment is highly variable and essentially without meaning.

Transitivity is also required of luminosity coefficients. When

$$(V_r + E_r) = (V_b + E_b) \tag{15.5}$$

(evoking red and blue sensations of equal brightness) and when

$$(V_y + E_y) = (V_g + E_g) \tag{15.6}$$

(evoking yellow and green sensations of equal brightness), then according to presumption, these brightness-equivalence pseudo-equations may be added physically to achieve a valid compound brightness equation:

$$(V_r + E_r) + (V_y + E_y) = (V_b + E_b) + (V_g + E_g). \tag{15.7}$$

The left field now evokes an orange sensation (red plus yellow stimuli), and the right field now evokes a bluish-green sensation (blue plus green stimuli). Although the hues are vastly different, both fields are predicted to evoke sensations with the same brightness. This additivity of luminance or brightness is usually termed Abney's Law (or, more accurately, the Abney-Festing Law): the luminance of a compounded color is the sum of the luminances of its components. Abney's Law is scattered through the literature, cited by MacAdam (1938), for example—but whence comes this law?

Abney's Law is named for Captain William de Wiveleslie Abney, a nineteenth-century Englishman who was graduated from the Royal Military Academy and stationed with Her Majesty's Royal Engineers in India, one of the outposts of the great British Empire. On his return to England in 1869, he became an instructor of military engineering, with an abiding interest in photography and astronomy. He was a fellow of the Royal Society of London, was awarded its distinguished Rumford Medal in 1882, and was knighted by Queen Victoria in 1900. He wrote many books on varied subjects, among them *Chemistry for Engineers, A Treatise on Photography* published in ten editions, and, strangely, *Thebes and Its Five Greater Temples.* Abney had a phenomenal output concerning luminosity and color vision, and his papers were published in distinguished journals. He wrote, for example, "Colour Photometry" (in three parts—the first reporting data confirming Abney's Law) with Major-General E. R. Festing (1886, 1888, 1892). The first paper reports empirical, experimental confirmation of that phenomenon now known as Abney's Law. In one example, the "illuminating values" of a red, green, and violet stimulus were 203.0, 38.5, and 8.5, respectively. The stimuli were mixed, and the "illuminating value" of the compound was 250.25—very close to the theoretical 250.00.

Abney (1900a, 1900b, 1910) wrote many papers on the general topic of luminance and brightness and two books on color vision emphasizing luminous efficiency (1895, 1913). With the exception of Abney's Law, his work had almost no influence on the subsequent development of colorimetry. Not even his luminosity data were accepted as reliable by subsequent colorimetrists. (See Troland, 1922.)

Perhaps Abney's Law should be called "Rood's Law." Abney was directly anticipated by Ogden N. Rood, a professor at Columbia University. Rood (1878) used Maxwell disks and calculated the luminances of test stimuli by matching their brightnesses to an achromatic standard. Rood found the luminance of a compounded color sensation to be the sum of the component luminances. He wrote: "This agreement proves the correctness of the photometric comparison, and also of Grassmann's assumption, that the total intensity [luminance] of the mixture of masses of differently colored light is equal to the sum of the intensities of the separate components, which appears not to have received prior experimental confirmation." Rood is best remembered for his charming books on color vision (1895).

Selig Hecht (1924) reported that the photopic and scotopic luminosity functions had a high correlation with the functions describing retinal pigment absorption at high and low light intensities. This was the beginning of an enormous amount of research, but that is not the concern of this work.

Brightness Matching Functions

By Abney's Law, luminance has a direct representation in pseudo-color-matching equations. Here is a pseudo-equation depicting the luminance of a sensation C compounded from three monochromatic components, B,G,R:

$$(\mathbf{L_C E_C}) \equiv (\mathbf{L_B E_B}) + (\mathbf{L_G E_G}) + (\mathbf{L_R E_R}). \qquad (15.8)$$

This pseudo-equation states that the left stimulus evokes a sensation with brightness equivalent to the combined brightnesses of the sensations evoked by three stimuli on the right. This is *not* a color-matching equation but rather a brightness-matching equation. This pseudo-equation for brightness will be of paramount importance in our discussion of color-matching functions.

Validity of Luminosity Coefficients

I make inquiry into the empirical properties of luminosity data and whether they in fact do predict brightness. By the tenets of the CIE in 1931, stimuli of equal luminance should evoke color sensations of equal brightness. By the teachings of the CIE in 1931, Abney's Law was valid within wide ranges of illumination, and the brightness of a color sensation was equivalent to the sum of the component brightnesses.

Luminosity, alas, is *not* a predictor of sensation brightness. David L. MacAdam (1950) reported a catastrophic empirical contradiction:

> It is easy to show that . . . [Abney's] law of additivity, which necessarily applies to luminance because of its definition, does not apply to brightness. When a white mixture was kept unchanged in the comparison half of the field, and the red component of an equally bright white mixture was reduced in intensity, the resulting color was a bluish-green, very obviously *brighter* than the white. By merely removing red light from a white mixture, a brighter color was produced [*sic*]. The luminance as well as the energy content was obviously decreased, because the luminous red component was removed without any compensating increase of the other components. In order to obtain the resulting saturated color with brightness just equal to white, it was also necessary to reduce the intensity of the green primary. This anomaly in particular and others like it forced themselves on the attention of the observers during the determinations of the constant hue loci. It cannot be attributed to a confusion of saturation with brightness. The observers tried without success to convince themselves that the bluish-green was darker than the white. They found that interpretation of their sensations quite unacceptable, and after keeping their attention directed to the saturation, they still judged the bluish-green to be brighter than the white from which it was derived by elimination of the red component.

Officially the CIE has now designated the word *luminance* as the function originally adopted in 1924, and *brightness* as another construct that might be predicted from empirical manipulation of the inviolate luminance function.

Peter F. Kaiser (1971) strenuously objected to the confusion engendered by the new official definitions of luminance and brightness. He wrote:

> The $V_{(k)}$ function is defined by the CIE and has its basis in the psychophysical data in the step by step brightness matching and minimum flicker methods [Gibson-Tyndall data]. These methods yielded a function with irregularities. The function adopted by the CIE had these irregularities mathematically smoothed. . . .
>
> I have been unable to find a definition of brightness in terms of the operation by which it is measured. . . . I propose the term be restricted to data collected by direct brightness measures. . . .
>
> At one time, the term luminance was called *photometric brightness*. Confusion prompted the CIE to adopt the term luminance [for this concept]. . . .
>
> An additional source of confusion stems from the CIE, which has given the terms *luminance* and *luminosity* different meanings. Luminance is

> defined physically [by an equation with defined elements], but luminosity is equivalent to brightness. . . . The unwary could easily fall in the trap of confusing luminance with brightness. If these two terms were equivalent, the additivity law (Abney's Law) should hold for additivity of brightness as well as additivity of luminances. It does not.
>
> . . . There are experiments in addition to those testing Abney's law that differentiate between luminance and brightness. Dresler (1953), Sanders and Wyszecki (1957), Wyszecki and Sanders (1957), and Kaiser and Wyszecki (1978) had a subject match a white for brightness with a chromatic one. If fields of equal luminance but differing in color were judged as equally bright, the ratios of luminance of the white to chromatic lights should of course equal 1.0. They were almost always greater than unity.
>
> . . . Since brightness and luminance data can differ, and both have a psychophysical basis, it seems reasonable to seek a physiological correlate based on these differences.

Breneman and Bartleson (1966) have indicated further displeasure with the concepts of luminosity and brightness, but based on semantic and philological arguments.

The official photopic visibility function is not the equivalent of a subjective brightness measure. The current literature is replete with titles that demonstrate this lack of equivalence. Here is a sampling: Burns, Smith, Pokorny, and Elsner (1982) wrote on the brightness of equal-luminance lights, Wyszecki (1967) wrote on a correlate for lightness in terms of luminous reflectance, Onley and Sternheim (1967) wrote on monochromatic stimuli of equal brightness but unequal luminance, Booker (1978) wrote on the luminance of equally bright colors, and Chapanis and Halsey (1955) wrote on the luminance of equally bright white lights.

It has now been nearly 75 years since the CIE adopted the official luminosity function. Researchers tenaciously hold to the function, presumably because it is described by the word *official*, and they strive to obtain a correlate of brightness by mathematical manipulation of the original official function. However, an infinity of other functions is available, unofficial as they should be, and color science may find it prudent to begin again.

The CIE Color-Matching Functions

The first color-matching functions were measured by Maxwell (1860) and later were recorded by König and Dieterici (1886). These data were modified by Weaver in accordance with supplemental data recorded by Abney (1913) and

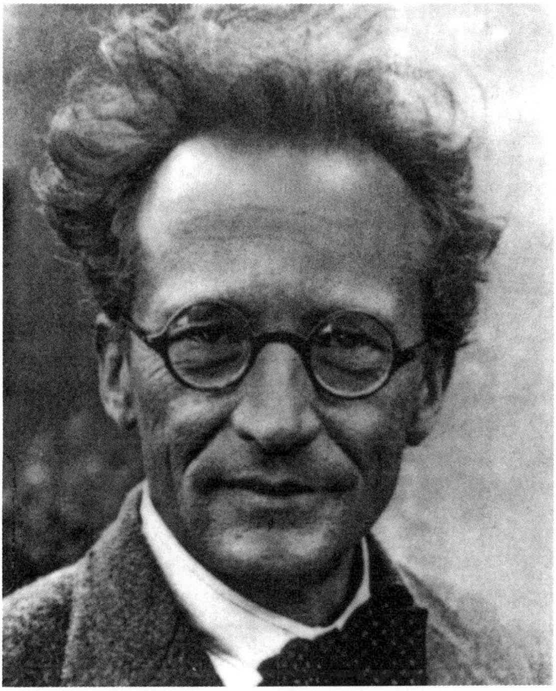

Erwin Schrödinger, 1887–1961

adopted as the official excitation curves of the Optical Society of America in 1922 (see Troland, 1922). By 1931, however, the art and science had advanced, and the CIE had two separate sets of color-matching data: the one by Guild (1924–1925a, 1924–1925b) and the other by Wright (1929, 1929–1930). When properly transformed, these data were closely coincident, and the CIE adopted an amalgam of these data to become the basic primaries for all future reference. But which transformation to which primaries was to become the recommended standard?

In 1931, the CIE declared its official luminosity function to be an official color-matching function; in adopting the standard primaries, the commission was greatly influenced by the theoretical writings of Nobel laureate Erwin Schrödinger (1920a, 1920b, 1925), who had analyzed a two-dimensional projection of an affine color space where distance and angle were meaningless. Schrödinger specifically warned against the introduction of Euclidean geometry in the study of color metrics, precisely the identical geometry successfully demonstrated by this book using the modern projection operator!

Within the projection, Schrödinger had proposed a single line, termed the *alychne* (Greek, "without light"), to specify theoretical color sensations lacking all "brightness." Two imaginary primaries resided on this line and represented theoretical color sensations without subjective "brightness." One imaginary primary resided outside the alychne and represented only subjective "brightness" in all color sensations. Thus, the CIE chose imaginary color stimuli: X and Z on the alychne (without subjective "brightness") and Y outside the alychne (with *all* subjective brightness.)

The luminosity function had already been adopted in 1924 and was not guaranteed to be a color-matching function by any a priori reasoning. Was it a color-matching function as Schrödinger had theorized, or wasn't it? Alas, it was not! In retrospect, it was highly improbable that the previously adopted luminosity function would coincide with a color-matching function from the Guild-Wright data. Nevertheless, the CIE proceeded to integrate the previously adopted luminance function into their conceptual model and deliberately changed the Guild-Wright color-matching data to achieve this accommodation. The CIE emerged with the now-famous X,Y,Z color-matching functions, where X and Z represented imaginary primaries that evoked color sensations without subjective "brightness" and Y (presumably) represented an imaginary primary that evoked color sensations with subjective "brightness" alone.

Shortly after this prescriptive CIE adoption, David L. MacAdam was a staff member in the Color Measurement Laboratory of the Massachusetts Institute of Technology. The staff, under the direction of A. C. Hardy (Hardy & MacAdam, 1936), published the definitive *Handbook of Colorimetry* designed to propagate the CIE model. It proclaimed, "In the case of the green paint selected for illustration, the tristimulus values that result from a computational procedure are X = 15.50, Y = 24.19, and Z = 22.64. . . . Thus the green . . . has a brightness [*sic*] relative to the white standard of 24.19%." More than a half-century later, MacAdam (1983) finally conceded the impropriety of this quotation: "Recent articles . . . indicate doubts concerning the validity of use of luminosity data in the 1931 CIE color-matching functions. Admittedly, the luminosity data used are incorrect."

In 1964, color-matching data collected by Stiles and Burch (1955) and tested by Wyszecki (Wyszecki & Wright, 1965) and others found no color-matching function to coincide with the luminance function adopted in 1931 but still official. Therefore, the CIE selected a function in the 1964 data with close similarity to the official function that has remained intact to this day.

This policy has been questioned. Based on his own intensive computational investigations, Oscar Estévez (1982) has written: "In no case can a photometric [luminosity] function be a linear combination of [color mixture functions], not even the function using the flicker method. Most important, the CIE $V_{(k)}$ is not itself a linear combination of [color mixture functions], and therefore the synthesized [1931] CIE [color mixture functions of Guild and Wright] must be in error, especially in the spectral region around 465 nm." These statements are supported by experimental evidence reported by Sperling (1958) and by Stiles and Burch (1955). Estévez concluded, "I suggest that, contrary to what was done in the 1931 Standard Observer, the disciplines of photometry and colorimetry be considered as two separate and distinct branches of visual science. It has long been known that the CIE 'luminance' concept is not adequate and that the problem is a serious one."

Besides the Y function, the CIE (1986) specified other properties to describe its official X,Y,Z functions. Of these, the principal requirement was all positive values—because of general fearfulness of negative and imaginary numbers and because negative numbers were difficult to manipulate with rotary calculators, before the advent of powerful computers. This requirement was met by selecting imaginary primaries outside the envelope of the fundamental vector configuration.

Great difficulties beset the CIE, principally because the colorimetric "equation," as outlined earlier in this work, was not well understood. This was acknowledged by W. D. Wright (1958) in every edition of his widely read book:

> This symbolism of this [colorimetric] equation calls for some comment, as we have to recognize that there are two distinct kinds of symbols, thus (C), (R), (G), and (B) have no numerical significance [*sic*!] but represent the quality of the lights that are being mixed or measured, whereas q,u,v,w represent the amounts of the lights involved in the observation. . . .
>
> As an added distinction between a colour-matching equation and a normal type of algebraic equation in which all the symbols can be multiplied, divided, etc., in the usual way, it has been agreed by the CIE (1955) that colour-matching equations can be written using the equivalent or identity, sign ≡ . . . . As a further reminder of the meaning of a colour equation, it is useful to adopt the custom of reading the equivalent sign as "matches" or "is matched by."

Wright then argued that equal "amounts" of blue, red, and green primaries are required to evoke a sensation matching a sensation evoked by an equal-energy stimulus, and therefore the tristimulus values were adjusted to be equal. As argued before in this work, equal quantities of the CIE B,G,R primaries are not, in fact, required.

My arguments may be reinforced. Suppose that Wright's "white" was composed of two complementaries, say blue and yellow. I cannot observe any blue or yellow in the "white" sensation. Shall I now conclude that the blue and yellow are present in equal amounts? By this argument, all complementary color stimuli, when added to evoke a white sensation, are present in equal "amounts." This is manifestly false.

When it is realized that tristimulus values are merely arbitrary dimensionless multipliers, we may understand that these multipliers may be added legitimately to other values in the same column of matrix **A**; this operation is performed when computing tristimulus values of heterochromatic color stimuli. However, these cannot be added to values in other columns of matrix **A**, and therefore the CIE trichromatic coefficients and chromaticity diagram are mathematically unacceptable.

During its long history, the CIE model has generated confusion between subjective psychological "brightness," once believed to be totally represented by the function (Y), and psychological "whiteness" or "chroma." Psychological "brightness" and psychological "whiteness" are very different entities. Nevertheless, the CIE has equated the Y luminance function with psychological "whiteness" and the luminosity function is often represented as a "vertical" axis on the CIE chromaticity diagram. The one-dimensional manifold of psychological black-gray-white color sensations without hue is described by the word *achromatic*, not *luminance*, and the two-dimensional manifold of the remaining two psychological color attributes is described by the obverse, *chromatic*.

The word *achromatic* does indeed refer to psychological color sensations that may vary only in the attribute of brightness, but every color sensation, even those with a large chromatic component, also vary in brightness. When I vary the brightness of a color sensation with achromatic and chromatic components, both the achromatic and chromatic components vary. In the current transmission of color television, the single signal destined for black-and-white monitors is termed *luminance*, and the other two signals destined for color monitors are termed *chrominance*. This nomenclature cannot be correct; if it were true, the chrominance signal alone (devoid of luminance) should be invisible on the receiving color monitor, but it *is* very visible, as may be demonstrated on almost all color television receivers.

The CIE model contrasts sharply with the orthonormal canonical representation of fundamental color space, where the first achromatic axis coincides with the vector representing an equal-energy white stimulus and sensation, and the remaining two axes define a true chromaticity plane, totally independent of each other and also independent of the achromatic component.

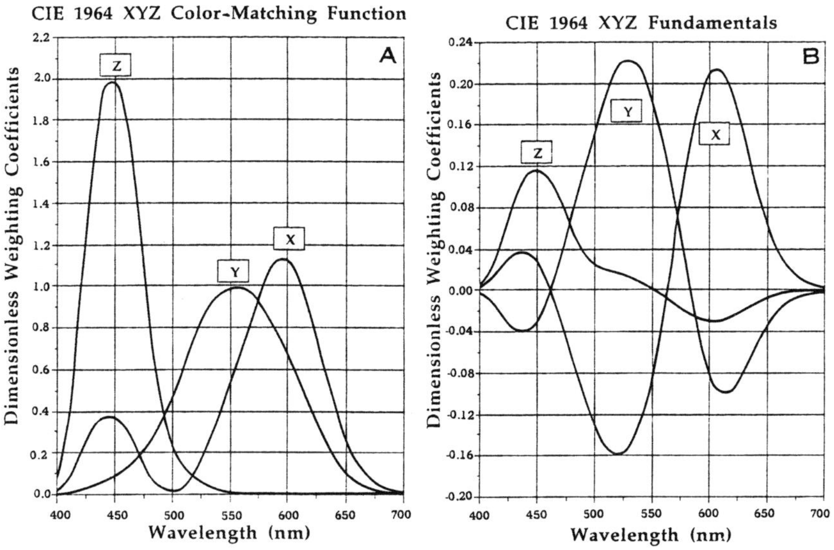

Figure 33. The marked difference between color-matching functions and fundamentals of primaries.

Fundamental Color Space and the CIE Primaries

Every color stimulus has a unique fundamental, scalar or vector, evoking a unique color sensation. This is true for monochromatic stimuli, and when the CIE specified its R,G,B primaries, it spoke of the primaries as monochromatic, effectively specifying these stimuli in terms of radiometric functions. However, the CIE gave only the color-matching functions when it introduced X,Y,Z. A critical question is this: What is a radiometric specification of X,Y,Z primaries—assuming that these imaginary primaries have the same properties as all other color stimuli? By my determinations, this salient question was not addressed during the many decades since the adoption by the CIE. Here, the question is answered by matrix-**R** operations. I derive the scalar and vector fundamentals of CIE primaries X,Y,Z.

Scalar Representation

Figure 33A depicts the three well-known color-matching functions, X,Y,Z, adopted by the CIE in 1964. The Y function is similar to the official luminance function adopted in 1931 and was presumed to be a measure of psychological brightness; this is not sustained by empirical evidence. The functions, as required, have all positive values.

A color-matching function is not a radiometric function, and it is generally not a fundamental of the primaries used in the color-matching experiment. With the introduction of matrix-**R** operations, I am able to compute the scalar fundamentals of these imaginary color stimuli (matrix **E**) from the color-matching functions provided by the CIE (matrix **A**). I use the usual $\mathbf{E} = \mathbf{A}(\mathbf{A}'\mathbf{A})^{-1}$. These fundamentals are illustrated in Figure 33B. The fundamentals are imaginary stimuli and cannot be produced physically; nevertheless, we may add residuals to produce imaginary metamers, exactly as I may add residuals to real fundamentals to produce new metamers. However, residuals added to imaginary fundamentals always result in imaginary metamers—and none may be reproduced physically; the imaginary metamers, however, remain theoretically interesting. (For an extended demonstration of these imaginary metamers, see Cohen, 1988.)

In all the years since the adoption of the CIE X,Y,Z color-matching functions, the CIE primaries could be specified only by their tristimulus values. With the advent of matrix-**R** operations, the scalar fundamentals of X, Y, and Z were calculated finally, as shown in Figure 33. For a vector fundamental representation of X,Y,Z, see Figure 36.

I explore additional meanings of valid color-matching equations. Figure 34A illustrates the radiometric function (matrix **N**) of Munsell 10BP 5/6 viewed under an equal-energy source; *B* is the fundamental of this stimulus obtained by matrix-**R** operations (matrix **N***), and *C* is its residual (matrix **B**). The tristimulus values are computed by the equation $\mathbf{Q} = \mathbf{A}'\mathbf{N}$; the tristimulus values are X = 22.4557, Y = 19.9364, and Z = 34.0206. Note that the tristimulus values are identical to those for the radiometric function and that the tristimulus values of the residual, as always, are 0, 0, 0.

The valid color-matching equation states that the color sensation evoked by matrix **N** or **N*** is matched by 22.4557 multiplied by the fundamental of primary X, plus 19.9364 multiplied by the fundamental of primary Y, and 34.0206 multiplied by the fundamental of the primary Z. The primaries X,Y,Z are primaries, and like all other color stimuli, real or imaginary, each has a unique scalar fundamental. The X,Y,Z fundamentals were shown in Figure 34B. Here these same fundamentals have been weighted by their respective tristimulus values (shown in Figure 34A and B), and appear in Figure 35A. The sum of these weighted fundamentals is shown in *B* and exactly duplicates the fundamental of the Munsell chip shown in Figure 34B. The sum of the X,Y,Z fundamentals, each arbitrarily weighted by .3333, is also shown in Figure 35B and is precisely the fundamental of an equal-energy stimulus as shown, for example,

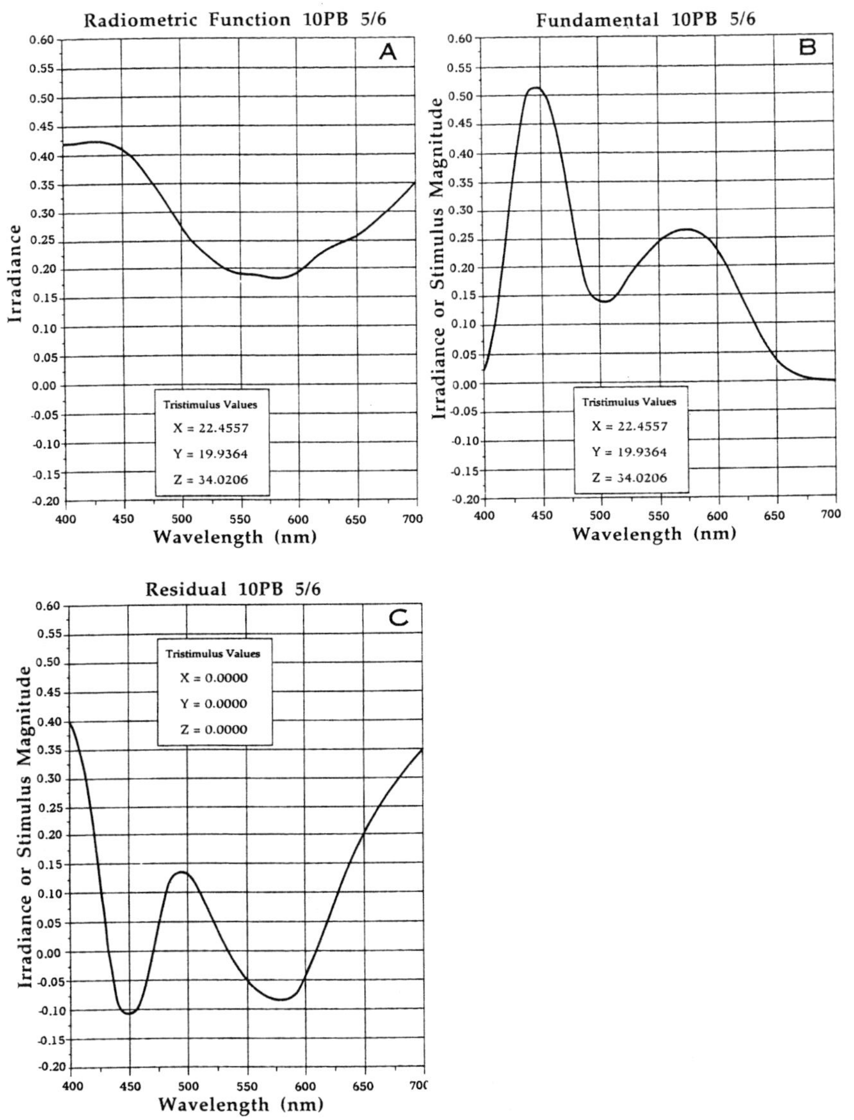

Figure 34. Radiometric function, fundamental, and residual of a Munsell chip.

in Figure 32E. Only the tristimulus values have been made to be equal; this does not mean that equal quantities of X, Y, and Z compound to a stimulus evoking the sensation of an equal-energy white. The areas under the function in Figure 34B are very different, and their vector fundamentals are of very different lengths; see Figure 36.

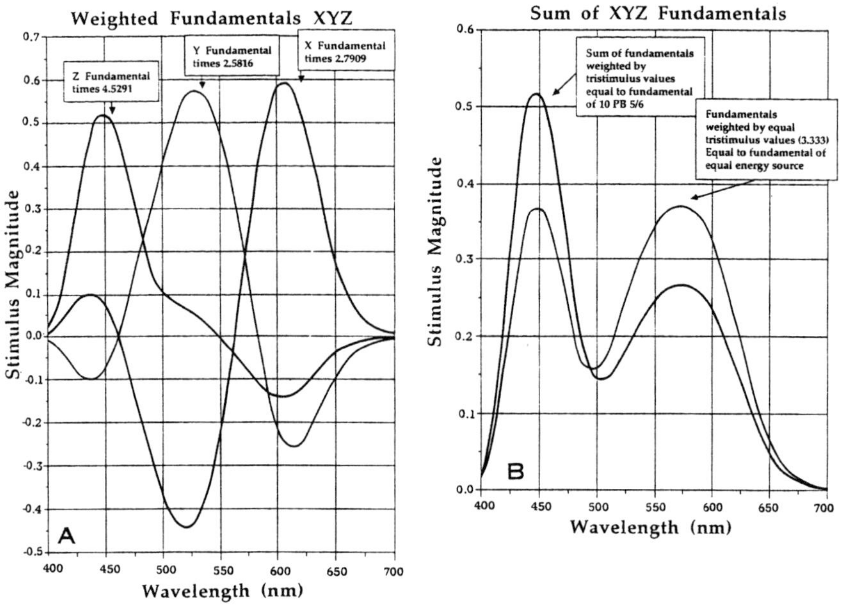

Figure 35. Weighted fundamentals and their sum.

Vector Representation

Given the 1964 R,G,B and X,Y,Z color-matching functions, I may immediately compute the position of the six vectors representing these primaries in fundamental color space. I need an arbitrary reference frame and have chosen a canonical orthonormal triplet of axes, with the first placed through the vector for the equal-energy source and the second and third orthogonal to each other and on a plane orthogonal to the first axis. The orthonormal frame is matrix **F**. Then, $\mathbf{F}'\mathbf{N}_p^*$ gives the coordinates, three for each primary.

The "butterfly wing" configuration of fundamental color space impedes viewing the simultaneous location of these six vectors. Nevertheless, they are depicted in the computer renditions in Figure 36A and B. The EE "white" axis is at the center of the configuration. In *B*, the view is not from the equal-energy axis but, rather, from the side of the equal-energy axis, as may be observed. Primaries B (444.4 nm), G (526.3 nm), and R (645.2 nm) are monochromatic stimuli and lie on the envelope of the configuration. Primaries X, Y, and Z are imaginary and lie external to the envelope of the fundamental color space. The primary X, unlike primaries Y and Z, does not correspond to a "supersaturated" monochromatic spectral stimulus. Notice that the X fundamental is outside

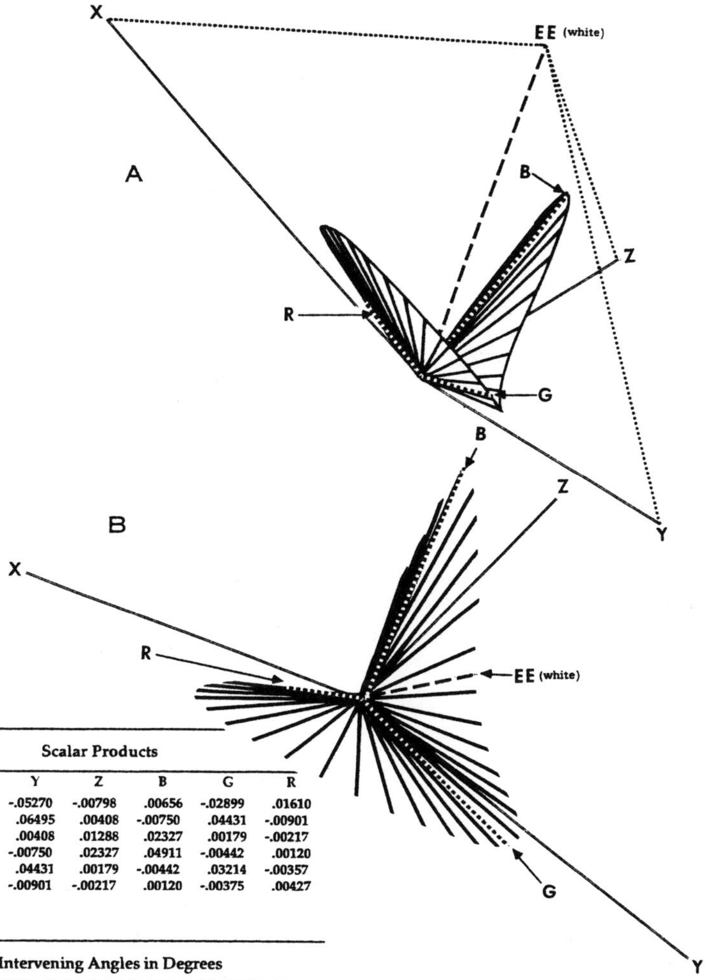

Scalar Products

Vector	X	Y	Z	B	G	R
X	.06835	-.05270	-.00798	.00656	-.02899	.01610
Y	-.05270	.06495	.00408	-.00750	.04431	-.00901
Z	-.00798	.00408	.01288	.02327	.00179	-.00217
B	.00656	-.00750	.02327	.04911	-.00442	.00120
G	-.02899	.04431	.00179	-.00442	.03214	-.00357
R	.01610	-.00901	-.00217	.00120	-.00375	.00427

Intervening Angles in Degrees

Vector	X	Y	Z	B	G	R
X		142.27	105.60	83.50	128.21	19.54
Y	142.27		81.89	97.63	14.08	122.73
Z	105.60	81.89		22.29	84.94	107.01
B	83.50	97.63	22.29		96.38	85.23
G	128.21	14.08	84.94	96.38		108.67
R	19.54	122.83	107.01	85.23	108.67	

C

Figure 36. Demonstration of the position of the X,Y,Z fundamentals.

the "purple" region (see *B*) of the configuration and not outside the red region.

Given the coordinates of the six vectors in an orthonormal reference frame, I have computed the scalar products and the angles between each pair of vectors. The results are the data shown in Figure 36C. The square roots of the diagonals are the vector lengths, and with respect to X,Y, and Z, represent the magnitudes of these primaries necessary to match the white sensation evoked by an equal-energy stimulus; note that the magnitudes are not equal.

Papers by Shapiro (1966) and by Ohyama, Yachida, Badique, Tsujiuchi, and Honda (1988) have independently attempted computation of these scalar products and angles for the 1931 X,Y,Z. Their results are virtually identical, reporting 40.5 degrees for X-Y, 84.3 degrees (Shapiro) and 85.3 degrees (Ohyama et al.) for Y-Z, and 75.3 degrees for X-Z. These values are incorrect because these researchers have used the color-matching functions in matrix **A** (highly variable under transformation) and not the primary fundamentals in matrix **E** (invariant under transformation). The variable color-matching functions are not the invariant fundamentals of the primaries in a color-matching equation. Using matrix **A**, all angles and vector lengths are different, even when only one primary is transformed. However, scalar products and angles computed from matrix **E** are invariably constant between the same vector pair—and vector lengths remain invariant.

I compute the coordinates of the vector fundamental for Munsell 10 PB 5/6. This vector is located by $\mathbf{S} = \mathbf{F}'\mathbf{N}^*_{\mathrm{m}}$, where $\mathbf{N}^*_{\mathrm{m}}$ is the scalar fundamental (or radiometric function) of the Munsell chip, matrix **F** is the canonical orthonormal frame, and 3×1 matrix **S** is the three coordinates of the $\mathbf{N}^*_{\mathrm{m}}$ in fundamental color space. The same vector is computed by multiplying the X,Y,Z fundamental vectors respectively by the chip's X,Y,Z tristimulus values. In every case, the X, Y, and Z vectors have been extended by the respective tristimulus values, all exceeding unity; the vector sum, or resultant, of these adjusted X,Y,Z vectors is the vector fundamental representing the Munsell chip BP 5/6. Note that the tristimulus values do not appear in fundamental color space; tristimulus values are merely dimensionless numbers modifying extant vectors within the fundamental color space.

Unification of the CIE 1931 and 1964 Standard Observers

Matrix-**R** operations suggest that the 1964 data may be reconciled with the 1931 data. Because the CIE has two standard observers, I have two matrix **R**'s: $\mathbf{R}_{31}$ and $\mathbf{R}_{64}$. These may be combined mathematically. I compute

$$\mathbf{R}_{31}\mathbf{R}_{64} = \mathbf{R}_{\mathrm{c}}, \tag{15.9}$$

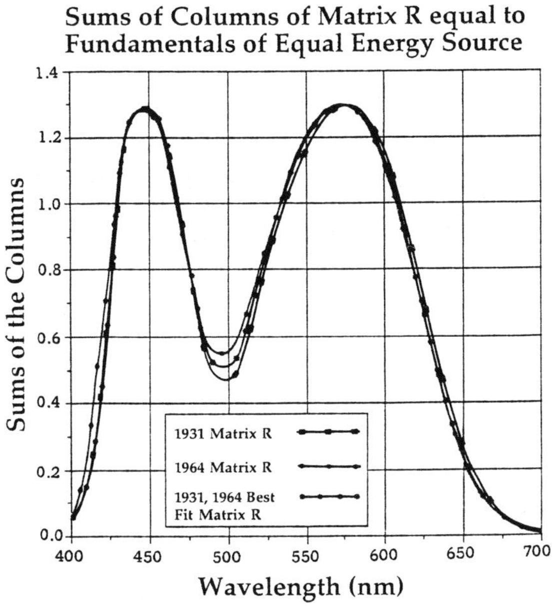

Figure 37. Matrix **R** operations permit interpretation of the CIE 1931 color-matching data and the CIE 1964 color-matching data.

where $\mathbf{R}_c$ is the best-fitting matrix **R** in both the 1931 fundamental color space and the 1964 fundamental color space. Ordering is irrelevant because the **R** matrices are symmetrical. The sums of the columns in each **R** matrix should give the two-peaked fundamental of an equal-energy stimulus. The three functions, shown in Figure 37, demonstrate the deviation of $\mathbf{R}_{31}$ and $\mathbf{R}_{64}$ from $\mathbf{R}_c$.

16

A Synthesis of Color Theories by Matrix-**R** Operations

Researchers in color science have hypothesized a large collection of color theories, each predicting the characteristics of the three visual "channels" presumed to transmit the color stimulus to the sensorium; the signal is interpreted as a color sensation. Almost invariably, these channels are described as three color-matching functions—in other times called *excitation curves*, *visual response functions*, or *distribution curves*. Three color-matching functions emerge from a color-matching experiment using three primary color stimuli, each primary with a unique fundamental with unique representation in fundamental color space.

Of color-matching functions Deane B. Judd (1930) has written: "It was indicated by Newton and subsequently verified by others, that this information, or mixture data, could be embodied in . . . additive distribution curves. From these curves directly can be obtained the prediction of whether two given stimuli of different spectral composition will appear matched or not. . . . It has long been recognized . . . that we can define an infinite number of sets of distribution curves representing the distribution of different sets of primary color processes which would equally well represent the same body of mixture data." Each theory, according to Judd and almost all other colorimetrists, is described by a unique triplet of color-matching functions.

Judd (1930) described two main types of color theories. Referring to three component theories, he wrote, "This theory was [first] briefly stated in 1807[?]

by Thomas Young, and was elaborated by Helmholtz about 50 years later. It assumed the existence of three independent response functions in the human eye: one predominantly sensitive to the long-wave light and yielding the response *red*; a second predominantly sensitive to the middle-wave light and yielding the response *green*, a third predominantly sensitive to short-wave light and yielding the response *violet*." Referring to opponent color theories, Judd wrote:

> This theory was [first] explained in detail by Ewald Hering in 1878. It is based on an analysis of sensations of color rather than on the stimuli required to evoke them. It assumes that there are six unitary colors (red, yellow, green, blue, white, and black), no one of which partakes of any other; that is, for example, yellow is a basic color in its own right, not the product of combining red with green.... This [accompanying retinal] activity is not found in six separate systems, but in three opposing pairs of processes: black-white, yellow-blue, and red-green. Black and white blend to produce gray, but equal amounts of yellow and blue, and of red and green cancel to zero.

These were the two main classes of color theories, but over the years, many variations have been proposed, each a linear transform from one triplet of color-matching functions to another triplet of color-matching functions. The functions were selected to embody the inherent characteristics of the theory; for example, one color-matching function of the Hering theory was presumed to have high positive loadings in the blue region of the spectrum and also high negative loadings in the yellow region of the spectrum; clearly the positive and negative signs may be interchanged.

After the CIE's adoption of the 1931 standard observer, it became fashionable to write a 3 × 3 matrix that would transform the color-matching data labeled X,Y,Z to the color-matching data of this or that color theory. Researchers proposed, arguably, that this personally designated triplet of color-matching functions was "best" because the triplet also helped to predict other aspects of color vision—as after-images, color contrast, color-blindness of various categories, and so forth. Because an infinite number of transformations were possible, in principle, a distinct color theory was available to every person on earth.

In 1951, after a decade of studying the color-matching functions of even the earliest theorists, Judd (1951) presented a concise table of 12 matrices for transformation from the CIE X,Y,Z to 12 specified color theories. When Burnham, Hanes, and Bartleson (1963) reproduced Judd's transformations, they added still another transformation matrix and theory, for a total of 13. I have arbitrarily selected these transformations and theories for analysis, but many other transformations appear in the literature (e.g., Pearson & Yule, 1973; Gibson, 1981; and Ohta, 1982).

In matrix-**R** notation, color-matching functions are a triplet comprising matrix **A**. From one matrix **A**, I may transform to another, and each new matrix **A** is a new color theory. All color-matching functions, however, are dimensionless; every transform brings a different row vector configuration, where distance and angle have no meaning.

Matrix-**R** operations have contributed many new constructs to color science; perhaps the most salient construct is the identification of the primary fundamentals in matrix **E** from the color-matching data in matrix **A**. I use the now-familiar $\mathbf{E} = (\mathbf{A}'\mathbf{A})^{-1}$. I know, of course, that the valid color-matching equation indicates that the rows of matrix **A** (tristimulus values) weight the columns of matrix **E**, and the weighted sum *is* the fundamental of the matched color stimulus. The concept of a fundamental, however, was not available to colorimetrists before the work of Günter Wyszecki and the advent of matrix-**R** operations.

I have applied matrix-**R** operations to Judd's 13 triplets of color-matching functions, each specifying an important color theory. In Judd's order, these are the 13 theories (14 are listed, but 2 are identical):

1. Young, three components
2. Helmholtz, three components
3. Late König (dominator-modulator)
4. Ladd-Franklin, three components, and early König (identical to Von Kries cone response, theory 6)
5. Hering, opponent colors
6. Von Kries, Schrödinger, zone or stage, cone response (identical to Ladd-Franklin, theory 4)
7. Von Kries, Schrödinger, zone or stage, optic nerve
8. Adams, zone or stage, cone pigments
9. Adams, zone or stage, cone response (CIE, X,Y,Z)
10. Adams, zone or stage, optic nerve
11. Müller, zone or stage, cone pigments
12. Müller, zone or stage, cone response
13. Müller, zone or stage, optic nerve
14. Hurvich-Jameson, receptor response

Tranformation matrices for theories 4 and 6 are identical. I shall use only Ladd-Franklin for simplicity. Also note that theory 8 is identical to the CIE. I shall use only the CIE for simplicity. Thus, we have 13 different theories, one identical to the CIE X,Y,Z.

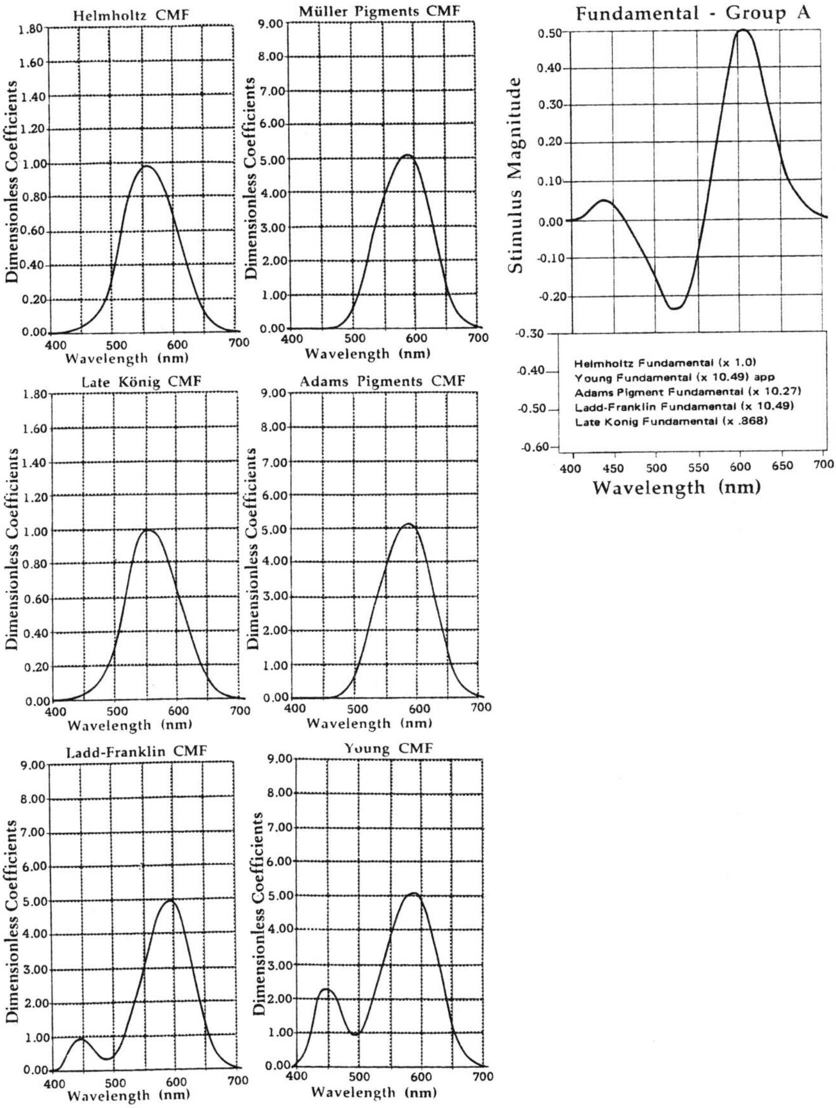

Figure 38. Demonstration 1 concerning color theories in the literature.

Because I have three color-matching functions and 13 different theories, we might expect 39 different primary fundamentals. Not at all!

We begin with the more important fundamentals.

As illustrated in Figure 38, the Helmholtz theory and late König theory have color-matching functions rather similar to the standard luminance func-

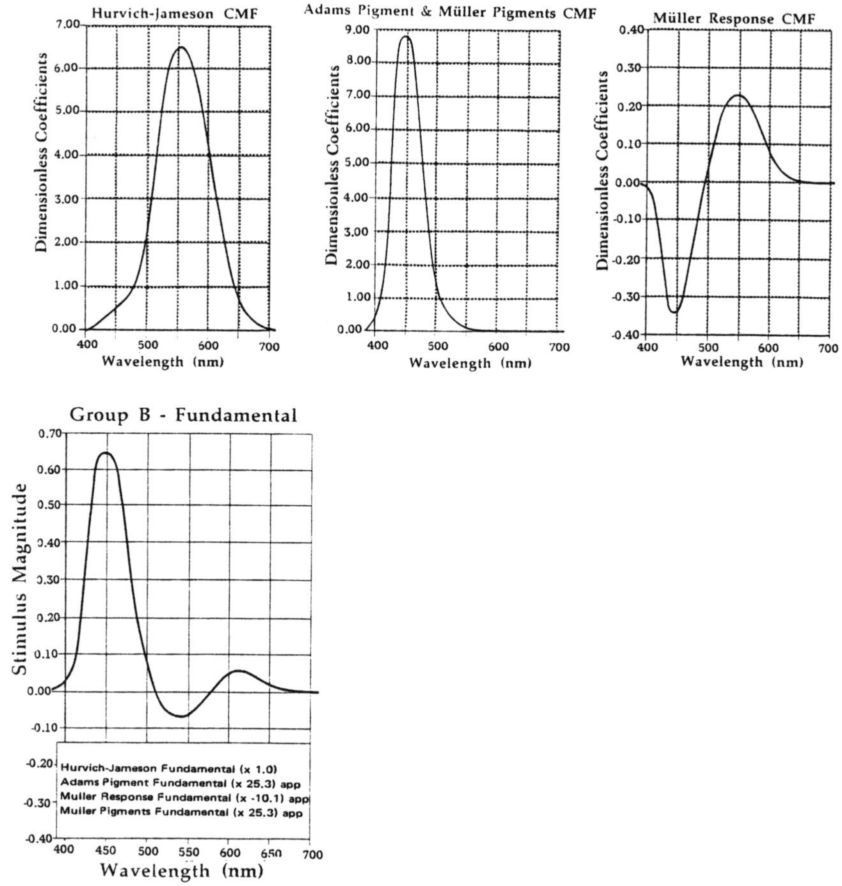

Figure 39. Demonstration 2 concerning color theories in the literature.

tion. Müller pigments and Adams pigments have identical color-matching functions, in this case, a luminance function believed to be characteristic of the color deficient. Ladd-Franklin and Young have similar yet different color-matching functions.

Despite this variance in color-matching functions, all six color-matching functions weight the identical primary fundamental (scaled with different constants). Although these six functions appear superficially to correspond to four different primaries, they in fact correspond to virtually the same primary.

With respect to Figure 39, observe a color-matching function of the Hurvich-Jameson theory, a color-matching function common to the Adams pigment and Müller pigment theory, and a still different color-matching function of the

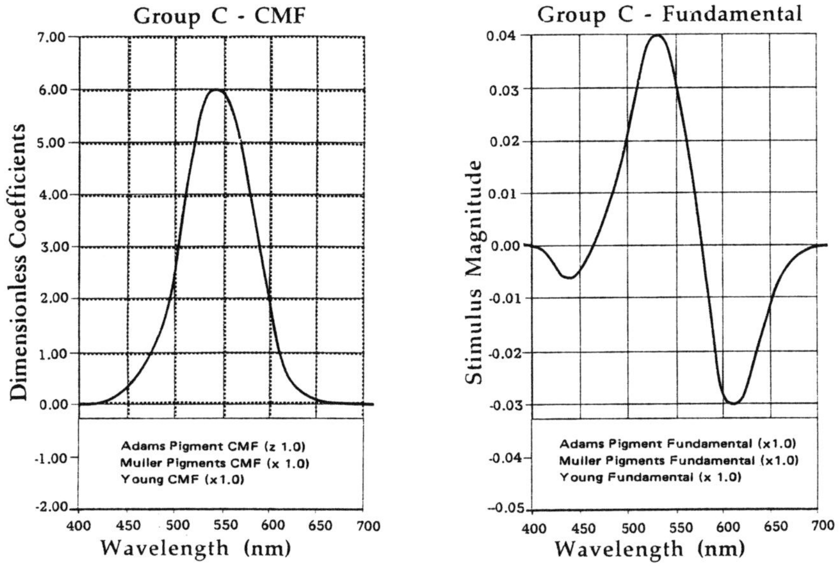

Figure 40. Demonstration 3 concerning color theories in the literature.

Müller response theory. All four, though very different from from each other, correspond closely to the fundamental.

In Figure 40, a color-matching function of the Adams pigment theory, the Müller pigment theory, and the Young theory are all identical, and all correspond to the same primary fundamental. In this case, the same color-matching functions correspond to the same primary fundamental.

In Figure 41, we have different color-matching functions from the CIE (X), Adams optic, and late König. Nevertheless, each corresponds to virtually the same primary fundamental.

In Figure 42, the standard luminance function (upper left) is common to the Müller response theory, the Hering theory, the Müller optic theory, the Adams optic theory, and the von Kries optic theory. This same luminance function corresponds to the fundamental of an equal-energy "white" stimulus (upper right), the same function that is the sum of the columns (or rows in matrix **R**) and has appeared so often in this book.

The luminance function is the Y function of the CIE standard color-matching functions X,Y,Z. Strangely, the luminance function in the CIE (lower left) does not correspond to the equal-energy white primary, but to its own peculiar primary fundamental (lower right). (The luminance function is often depicted as a vertical axis over the equal-energy point on the CIE chromaticity diagram;

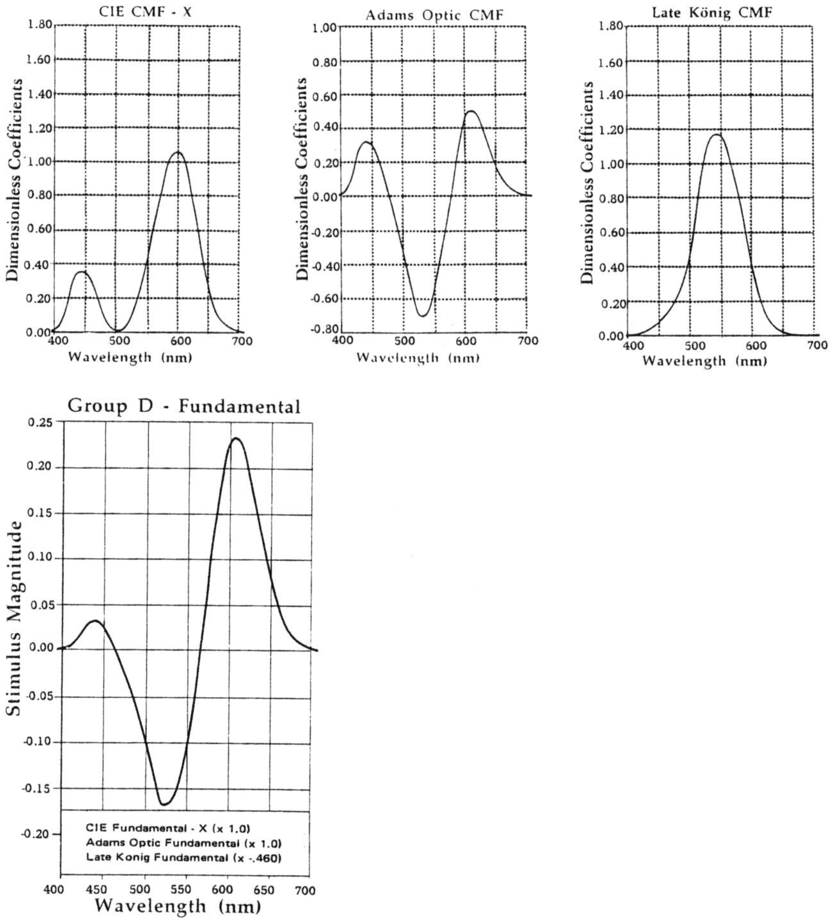

Figure 41. Demonstration 4 concerning color theories in the literature.

however, the luminance function (Y) in the CIE X,Y,Z color-matching functions does not correspond to the equal-energy fundamental.)

Finally, Figure 43 depicts color-matching functions from the CIE (Z), the Adams optic theory, the Müller optic theory, and the Hering optic theory. The last three seem closely related, the Adams optic being the mathematical complement of Hering color-matching functions. However, they are very different from the CIE Z color-matching functions. Nevertheless, all four color-matching functions have virtually the same primary fundamental.

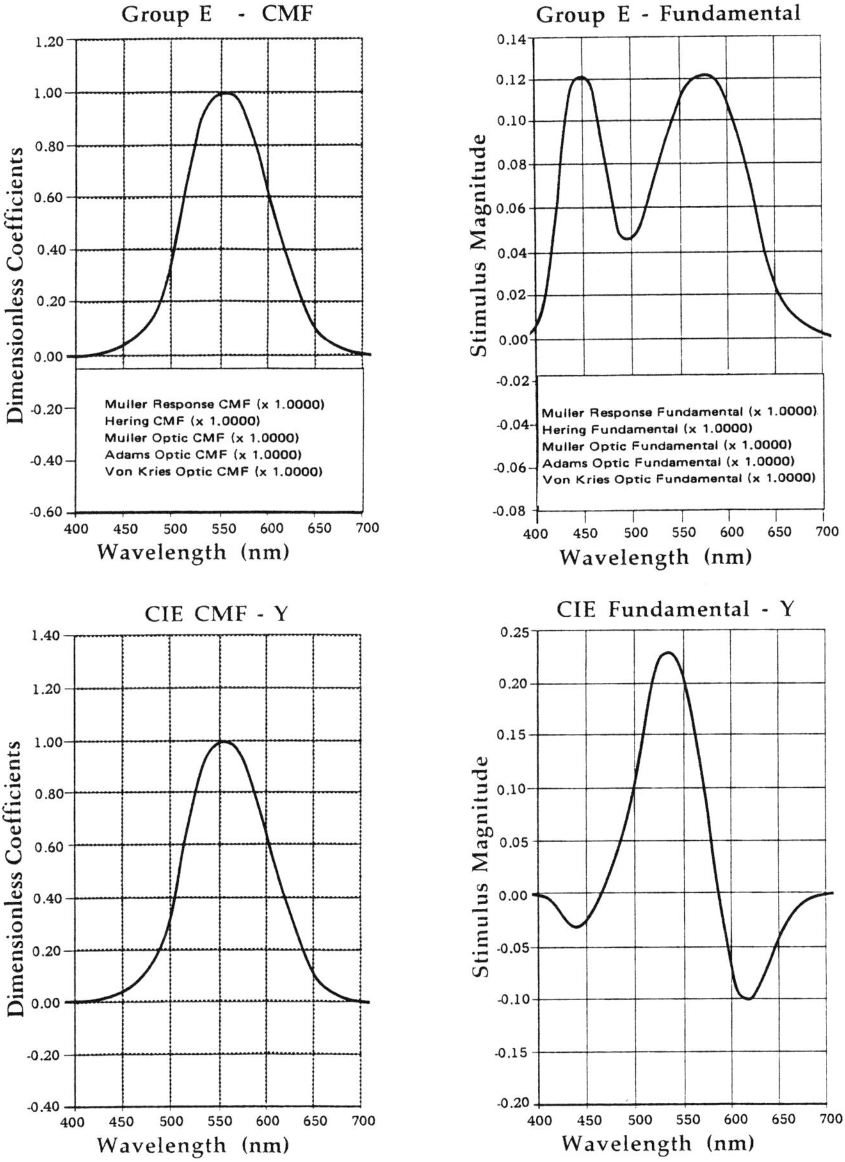

Figure 42. Demonstration 5 concerning color theories in the literature.

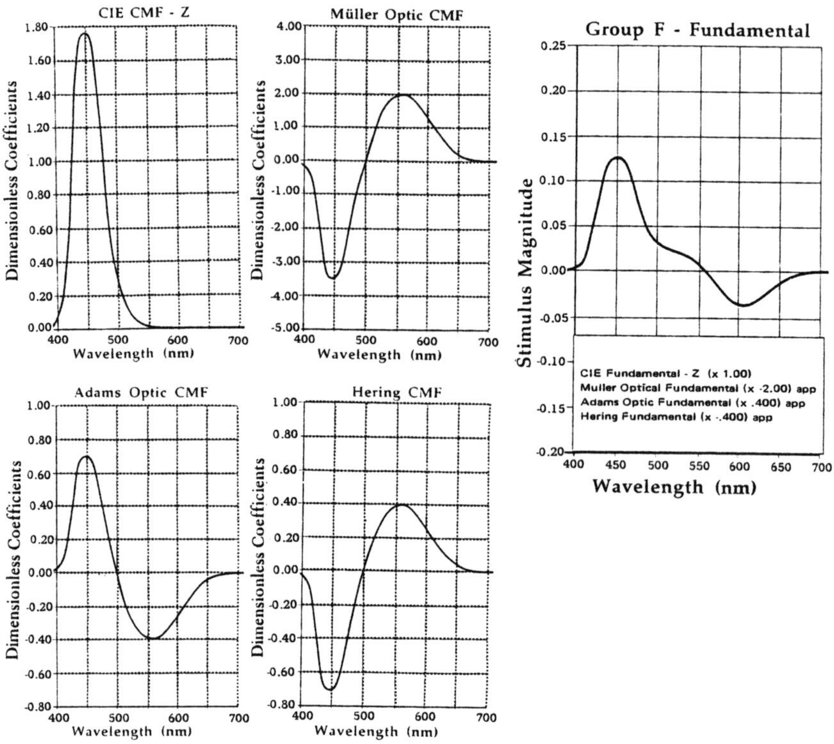

Figure 43. Demonstration 6 concerning color theories in the literature.

After these several telling representations of color-matching functions and their corresponding fundamentals, I insist that color theories cannot be defined by a triplet of color-matching functions. The reason is found in the prior discussions in this book to the effect that color-matching functions are highly variable and generally change when any one primary is changed. Color-matching functions are almost unpredictable. For example, many of the above theories transformed the Y luminance function intact to its preferred triplet of color-matching functions. After transformation, the luminance function is sometimes found to have an equal-energy "white" primary—but not invariably. In the CIE system itself, the Y color-matching function does not correspond to an equal-energy "white" primary. Color-matching functions are always in flux, in contrast to the fundamentals that remain fixed and invariable. In the selection of primaries for color theories, I recommend that the selection be made from fundamental vectors in fundamental color space.

With respect to several functions shown in Figure 42, Judd has made the luminance function correspond to and weight the fundamental of an equal-energy "white." At first I was astonished at the insight required to bring this about. However, I conjecture that this was accomplished by use of the theorem developed earlier in this book, that the sums of the tristimulus values (in matrix **A**) are the weights on the fundamentals, so that the sum of the weighted fundamentals *is* the fundamental of an equal-energy "white" stimulus.

To test its application to Judd's transformations, I have studied the color-matching functions and fundamentals for all theories discussed by Judd and also the theory of Hurvich and Jameson. For the Hering theory where the third color-matching function is the luminance function, the sums of the three color-matching functions are −.0021, .0035, and 10.6857. The first two sums are effectively zero. When the fundamentals are weighted by these tristimulus values, the first two columns will equal zero, and the luminance column (with the only nonzero tristimulus value) will weight the fundamental of the equal-energy "white" stimulus, as required by the theorem.

Judd's task, then, was to make the sums of the first two color-matching functions respectively equal zero. Without matrix-**R** operations, I can surmise only a trial-and-error technique, where the third color-matching function was held constant and the first two rotated until, finally, a position was found where the sums approached zero—as is required by the Hering theory itself.

I may, of course, plot the fundamentals of each theory in fundamental color space. I have done so for several theories and report that the triplet of fundamentals may occupy some very peculiar positions.

I have computed the angles between each triplet of fundamentals defined by each theory. None is orthogonal, not even tolerably so. I use the Hering theory as an example. Here are the scalar products between the fundamentals (matrix $\mathbf{M}^{-1}$):

$$\begin{bmatrix} .3114 & -.0559 & .1725 \\ -.0559 & .2889 & -.1441 \\ .1725 & .1441 & .4476 \end{bmatrix}.$$

The square roots of the diagonals are the respective vector lengths of the primaries. Here is the product of the vector lengths:

$$\begin{bmatrix} .3114 & .3000 & .3734 \\ .3000 & .2889 & .3596 \\ .3734 & .3596 & .4476 \end{bmatrix}.$$

Dividing corresponding elements in the two matrices, I obtain the direction cosines:

$$\begin{bmatrix} 1.0000 & -.1865 & -.4621 \\ -.1865 & 1.0000 & -.4007 \\ -.4621 & -.4007 & 1.0000 \end{bmatrix}.$$

And I compute the angles:

$$\begin{bmatrix} 0.00^\circ & 100.74^\circ & 117.52^\circ \\ 100.74^\circ & 0.00^\circ & 113.62^\circ \\ 117.52^\circ & 13.62^\circ & 0.00^\circ \end{bmatrix}.$$

An orthonormal frame in fundamental color space is probably the best specification for the transmission characteristics of the visual system. The problems of the theories here discussed derive from the instability of an **A** matrix. The rows of the tricolor functions would then weight the columns of the tricolor functions, and the inherent instability of **A** matrices would not exist.

Appendix A: BASIC Program for Orthogonalization of Fundamentals by the Gram-Schmidt Procedure

The program is written for matrices with seven rows; this may be changed to any *k*.

```
100 REM PLACE FUNDAMENTALS IN MAT D. EXAMPLE USES 7 ROWS.
SUBSTITUTE ANY
110 REM NUMBER. FIRST COLUMN OF OUTPUT WILL BE PROPORTIONAL TO
120 REM  COLUMN OF INPUT.
130 DATA x,x,x
140 DATA x,x,x
150 DATA x,x,x
160 DATA x,x,x
170 DATA x,x,x
180 DATA x,x,x
190 DATA x,x,x
200 LET AAA1$ = "- - -.####  - - -.####  - - -.####"
210 DIM D(7,3),X(7,3),X1(7,3)
220 MAT READ D
230 PRINT "INSERT fundamentals IN MATRIX D"
240 PRINT
250 PRINT "THESE ARE THE ORIGINAL INDEPENDENT VECTORS"
260 MAT PRINT USING AAA1$: D
270 LET I = 0
280 FOR I = 1 TO 7
290 LET X(I,1) = D(I,1)
300 NEXT I
310 PRINT "MATRIX K (1,3) HOLDS COEFFICIENTS AT EACH STAGE OF
GRAM-SCHMIDT."
320 DIM S(2,3)
330 DIM K(1,3)
340 PRINT
350 PRINT
360 DIM I9(3,3)
370 DIM T(3,3)
380 PRINT "CALCULATION OF X(I,1)"
390 LET I = 0
400 FOR I = 1 TO 7
410 LET X(I,1) = D(I,1)
420 NEXT I
430 PRINT "CALCULATION OF X(I,2)"
```

```
440 LET I = 0
450 FOR I = 1 TO 7
460 LET S(1,1) = S(1,1) + (X(I,1)^2)
470 NEXT I
480 LET S(2,1) = S(1,1)^.5
490 LET I = 0
500 FOR I = 1 TO 7
510 LET K(1,1) = K(1,1) + D(I,2)*X(I,1)
520 NEXT I
530 LET K(1,1) = K(1,1)/S(1,1)
540 LET I = 0
550 FOR I = 1 TO 7
560 LET X(I,2) = D(I,2) - K(1,1)*X(I,1)
570 NEXT I
580 MAT K = ZER
590 REM CALCULATION OF X(I,3)
600 LET I = 0
610 FOR I = 1 TO 7
620 LET S(1,2) = S(1,2) + (X(I,2)^2)
630 NEXT I
640 LET S(2,2) = S(1,2)^.5
650 LET I = 0
660 FOR I = 1 TO 7
670 LET K(1,2) = K(1,2) + D(I,3)*X(I,2)
680 NEXT I
690 LET K(1,2) = K(1,2)/S(1,2)
700 LET I = 0
710 FOR I = 1 TO 7
720 LET K(1,1) = K(1,1) + D(I,3)*X(I,1)
730 NEXT I
740 LET K(1,1) = K(1,1)/S(1,1)
750 LET I = 0
760 FOR I = 1 TO 7
770 LET X(I,3) = D(I,3) - K(1,2)*X(I,2) - K(1,1)*X(I,1)
780 NEXT I
790 PRINT
800 PRINT "THESE ARE ORTHOGONAL FUNCTIONS, NOT ORTHONORMAL"
810 PRINT
820 MAT PRINT USING AAA1$: X
830 MAT K = ZER
840 LET I = 0
850 FOR I = 1 TO 7
860 LET S(1,3) = S(1,3) + (X(I,3)^2)
870 NEXT I
880 LET S(2,3) = S(1,3)^.5
890 REM COMPUTATION OF ORTHONORMAL FUNCTIONS X1(1-3)
900 LET I = 0
```

```
910 FOR I = 1 TO 7
920 FOR J = 1 TO 3
930 LET X1(I,J) = X(I,J)/S(2,J)
940 NEXT J
950 NEXT I
960 MAT T = TRN(X1)
970 MAT I9 = T*X1
980 DIM I8(3,3)
990 PRINT "HERE ARE THE ORTHONORMAL FUNCTIONS"
1000 PRINT
1010 MAT PRINT USING AAA1$: X1
1020 PRINT
1030 PRINT "This is X1'X1. Should be an identity matrix"
1040 PRINT
1050 MAT PRINT USING AAA1$: I9
1060 PRINT "WE COMPUTE R"
1070 DIM R(7,7)
1080 MAT R = X1*T
1090 LET A$ = "- -.#### - -.#### - -.#### - -.#### - -.#### - -.#### - -.####"
1100 MAT PRINT USING A$: R
1110 END
```

Appendix B: BASIC Program for Calculating an **F** Matrix Using Matrix-**R** Operations

The method depends on factoring an **M** matrix into matrix **G**, such that **GG′** = **M**. Factoring is by the so-called diagonal method. The immediate input is an **E** matrix, three independent fundamentals. However, if the first column of the output matrix is to be proportional to the first column of the input matrix **E**, then the method requires a secondary input—matrix **A** computed from matrix **E**. Then the first column of the output matrix is proportional to the first column in matrix **E**, as in the Gram-Schmidt procedure. This method is powerful and far more efficient than Gram-Schmidt. The program is written for **E** and **A** matrices with seven rows, but this may be changed to any *k*. The method may also be extended to any number of dimensions, as is true for Gram-Schmidt.

```
100 PRINT “ORTHOGONALIZATION BY MATRIX-R PROCEDURES”
110 PRINT “ENTER DATA FOR THREE FUNDAMENTALS, MATRIX E”
120 PRINT “TRANSFORM TO MATRIX A, AND OPERATE ON MATRIX A”
130 PRINT “FIRST COLUMN OF OUTPUT WILL BE PROPORTIONAL TO FIRST
COLUMN IN”
140 PRINT “MATRIX E, NOT MATRIX A”
150 PRINT “THESE ARE DATA FOR AN E MATRIX”
160 DATA x,x,x
170 DATA x,x,x
180 DATA x,x,x
190 DATA x,x,x
200 DATA x,x,x
210 DATA x,x,x
220 DATA x,x,x
230 DIM E(7,3)
240 MAT READ E
250 LET A$ = “- - -.####  - - -.####  - - -.####”
260 PRINT
270 PRINT “THIS IS MAT E, THREE INDEPENDENT FUNDAMENTALS”
280 PRINT
290 MAT PRINT USING A$: E
300 PRINT
310 DIM E1(3,3),E2(3,3),A(7,3), M(3,3), M1(3,3)
320 MAT E1 = TRN (E)
330 MAT M = E1*E
340 MAT M1 = INV(M)
350 MAT A = E*M1
```

```
360 DIM R(7,7)
370 MAT R = A*E1
380 PRINT
390 PRINT "THIS IS THE M INVERSE MATRIX COMPUTED FROM MATRIX A"
400 PRINT
410 MAT PRINT USING A$: M
420 PRINT
430 PRINT
440 PRINT "THIS IS MAT A"
450 PRINT
460 MAT PRINT USING A$: A
470 PRINT
480 PRINT "AND THIS IS MAT R"
490 LET A1$ = " - -.#### - -.#### - -.#### - -.#### - -.#### - -.#### - -.####"
500 PRINT
510 MAT PRINT USING A1$: R
520 PRINT
530 PRINT
540 PRINT "WE ORTHOGONALIZE BY FACTORING THE M MATRIX TO GET G"
550 PRINT
560 DIM G(3,3)
570 LET G(1,1) = M(1,1)^.5
580 LET G(2,1) = M(2,1)/G(1,1)
590 LET G(3,1) = M(3,1)/G(1,1)
600 LET G(2,2) = (M(2,2) - G(2,1)^2)^.5
610 LET G(3,2) = (M(3,2) - G(2,1)*G(3,1))/G(2,2)
620 LET G(3,3) = (M(3,3) - G(3,1)^2 - G(3,2)^2)^.5
630 PRINT
640 PRINT "THIS IS MAT G"
650 PRINT
660 MAT PRINT USING A$: G
670 PRINT
680 PRINT
690 PRINT "MULTIPLY G TIMES G', SHOULD GET M"
700 DIM G1(3,3), M9(3,3), M8(3,3)
710 MAT G1 = TRN (G)
720 MAT M9 = G*G1
730 PRINT
740 MAT PRINT USING A$: M9
750 PRINT
760 PRINT "THIS IS G'G"
770 PRINT
780 MAT M8 = G1*G
790 MAT PRINT USING A$: M8
800 PRINT
810 PRINT "MULTIPLY AG, TO GET MATRIX F."
820 PRINT
```

```
830 DIM F8(7,3)
840 MAT F8 = A*G
850 PRINT
860 MAT PRINT USING A$: F8
870 DIM R2(7,7)
880 DIM F7(3,7)
890 MAT F7 = TRN (F8)
900 MAT R2 = F8*F7
910 PRINT
920 PRINT "THIS IS R BY FF'"
930 PRINT
940 MAT PRINT USING A1$: R2
950 PRINT
960 PRINT "COMPUTE AE'. SHOULD BE AN R MATRIX"
970 DIM R3(7,7)
980 MAT R3 = A*E1
990 PRINT
1000 MAT PRINT USING A1$: R3
1010 PRINT
1020 PRINT "WE COMPUTE F'F. SHOULD BE AN IDENTITY MATRIX"
1030 PRINT
1040 DIM I5(3,3)
1050 MAT I5 = F7*F8
1060 PRINT
1070 MAT PRINT USING A$: I5
1080 PRINT
1090 END
```

Appendix C: BASIC Program for Converting Scalar Products, as in **M** or **R** Matrices, to Lengths and Angular Separations Between Vectors

This program is written for six row vectors, but this may be converted to any *k*.

```
100 DATA x,x,x
110 DATA x,x,x
120 DATA x,x,x
130 DATA x,x,x
140 DATA x,x,x
150 DATA x,x,x
160 REM Mat F3 IS 6,3. EACH ROW SPECIFIES THE COORDINATES OF ONE
VECTOR
170 REM (IN FUNDAMENTAL COLOR SPACE). THE COORDINATES REFER TO
180 REM ONE ORTHONORMAL FRAME, F3.
190 REM BEGIN BY INSERTING MAT F3
200 REM
210 DIM F3(6,3), F4(3,6)
220 MAT READ F3
230 LET A$ = "- -.####  - -.####   - -.####"
240 LET A1$ = "- - - -.####  - - - -.####   - - - -.####  - - - -.####  - - - -.####  - - - -.####"
250 PRINT
260 PRINT "THESE ARE THE COORDINATES IN FUNDAMENTAL COLOR
SPACE"
270 PRINT
280 MAT PRINT USING A$: F3
290 PRINT
300 MAT F4 = TRN (F3)
310 DIM M1(6,6)
320 MAT M1 = F3*F4
330 DIM M2(6,6)
340 DIM Y(6,1)
350 DIM Y1(1,6)
360 LET I = 0
370 FOR I = 1 TO 6
380 LET Y(I,1) = M1(I,I)^.5
390 NEXT I
400 MAT Y1 = TRN (Y)
410 PRINT
420 PRINT "THESE ARE THE VECTOR LENGTHS OF THE VECTORS, THE"
430 PRINT "ROOTS OF DIAGONALS IN MATRIX M1"
```

```
440 PRINT
450 MAT PRINT USING A$: Y
460 PRINT
470 DIM S(6,6)
480 MAT S = Y*Y1
490 PRINT
500 PRINT "THESE ARE THE PRODUCTS OF THE VECTOR LENGTHS"
510 PRINT
520 MAT PRINT USING A1$: S
530 PRINT
540 DIM M3(6,6)
550 LET I = 0
560 LET J = 0
570 FOR I = 1 TO 6
580 FOR J = 1 TO 6
590 LET M3(I,J) = (M1(I,J))/(S(I,J))
600 NEXT J
610 NEXT I
620 PRINT
630 PRINT "THESE ARE THE DIRECTION COSINES BETWEEN THE VECTORS"
640 PRINT
650 MAT PRINT USING A1$:  M3
660 PRINT
670 DIM M6(6,6)
680 PRINT "THESE ARE THE ANGLES BETWEEN THE VECTORS,"
690 PRINT "IN DEGREES"
700 LET I = 0
710 LET J = 0
720 FOR I = 1 TO 6
730 FOR J = 1 TO 6
740 IF I=J THEN 760
750 IF I <> J THEN 770
760 LET M6(I,J) = 0
770 NEXT J
780 NEXT I
790 LET I = 0
800 LET J = 0
810 FOR I = 1 TO 6
820 FOR J = 1 TO 6
830 IF I = J THEN 860
840 IF I<>J THEN 850
850 LET M6(I,J) = ACOS (M3(I,J))
860 NEXT J
870 NEXT I
880 PRINT
890 MAT PRINT USING A1$: M6
900 PRINT
910 END
```

References

Abney, W. de W. (1895). *Colour vision, being the Tyndall lectures delivered in 1894 at the Royal Institution.* London: Sampson Low, Marston.

Abney, W. de W. (1900a). The colour sensations in terms of luminosity. *Philosophical Transactions of the Royal Society of London, 193A*, 259–287.

Abney, W. de W. (1900b). On the estimation of luminosity of coloured surfaces used for colour disks. *Proceedings of the Royal Society of London, 67*, 118–123.

Abney, W. de W. (1910). On the extinction of colour by reduction of luminosity. *Proceedings of the Royal Society of London, 83A*, 290–297.

Abney, W. de W. (1913). *Researches in colour vision and the trichromatic theory.* New York: Longman Green.

Abney, W. de W., & Festing, R. E. (1886). Colour photometry. *Philosophical Transactions of the Royal Society of London, 177*, 423–442.

Abney, W. de W., & Festing, R. E. (1888). Colour photometry. Part II. The measurement of reflected colours. *Philosophical Transactions of the Royal Society of London, 179A*, 547–570.

Abney, W. de W., & Festing, R. E. (1892). Colour photometry. Part III. Measurement of luminosity. *Philosophical Transactions of the Royal Society of London, 183A*, 531–565.

Albert, A. (1972). *Regression and the Moore-Penrose pseudoinverse.* New York: Academic Press.

d'Alembert, J. Lerond. (1751–1765). Couleur. In D. Diderot & J. L. d'Alembert (eds.), *Encyclopédie, ou Dictionnaire Raisonné des Sciences, des Arts, et des Métiers.* (17 vols plus plates.) Paris: Briasson, David, Lebreton, Durand.

d'Alembert, J. Lerond. (1761–1780). *Opuscules mathématiques; ou, Mémoires sur différens sujets de géométrie, de méchanique, d'optique, d'astronomie*, &c. (8 vols.) Paris: David.

d'Alembert, J. Lerond. (1789–1797). *Dictionnaire encyclopédique des mathématiques*, par mm. d'Alembert, l'abbé Bossut, De la Lande, le marquis de Condorcet, &c. Paris: Panckoucke. 3 vols. and atlas.

Allen, E. (1966a). Basic equations used in computer color matching. *Journal of the Optical Society of America, 56*, 1256–1259.

Allen, E. (1966b). Matrix algebra for colorimetrists. *Color Engineering, 4*(4), 24–31.

Allen, E. (1969). Some new advances in the study of metamerism. *Color Engineering, 7*, 35–40.

Amir-Moez, A. R., & Newman, T. G. (1970). Geometry of generalized inverses. *Mathematics Magazine, 43*, 33–36.

Arago, M. (1871). Eulogy on Thomas Young. In *Annual report of the Board of Regents of the Smithsonian Institution* (pp. 111–144). Washington, DC: Author.

Ayre, H. G., Stephens, R., & Mock, G. D. (1967). *Analytic geometry: Two and three dimensions.* Princeton, NJ: Van Nostrand.

Bailey, F. H., & Woods, F. S. (1898). *Plane and solid analytic geometry.* New York: Ginn.

Barnes, N. F. (1939). Color characteristics of artists' pigments. *Journal of the Optical Society of America, 29,* 208–214.

Ben-Israel, A., & Greville, T. N. E. (1975). *Generalized inverses: Theory and applications.* New York: Wiley-Interscience.

Berlin, B., & Kay, P. (1969). *Basic color terms: Their universality and growth.* Berkeley: University of California Press.

Berns, R. S., & Kuehni, R. G. (1990). What determines crossover wavelengths of metameric pairs with three crossovers. *Color Research and Application, 15,* 23–28.

Bingley, F. J. (1954). Transformations of chromaticity diagrams. *Journal of the Optical Society of America, 44,* 109–112.

Bird, G. R., & Jones, R. C. (1965). Estimation of the human cone pigments. *Journal of the Optical Society of America, 55,* 1686–1691.

Birren, F. (1979). Color identification and nomenclature: A history. *Color Research and Application, 4,* 14–18.

Birren, F. (1981). J. C. LeBlon, discoverer and developer of the red-yellow-blue principle of color printing and color mixture. *Color Research and Application, 6,* 85–92.

Bochner, S. (1962). The role of mathematics in the rise of mechanics. *American Scientist, 50,* 294–311.

Booker, R. I. (1978). Luminance-brightness comparisons of LED alpha-numeric sources at suprathreshold level. *Journal of the Optical Society of America, 68,* 949–952.

Boring, E. G. (1921). The stimulus-error. *American Journal of Psychology, 32,* 449–471.

Boring, E. G. (1929). *A history of experimental psychology.* New York: Century.

Born, M. (1953). Physical reality. *Philosophical Quarterly, 3,* 139–149.

Bornstein, M. C. (1973). Color vision and color naming: A psychophysiological hypothesis of cultural differences. *Psychological Bulletin, 80,* 257–285.

Bouguer, P. (1760). Traité d'optique sur la gradation de la lumière. Paris: H. L. Guérin, L. F. Delatour. (English translation by W. E. K. Middleton. *Optical treatise on the gradation of light.* Toronto: Univ. of Toronto Press, 1961.)

Boullion, T. L., & Odell, P. L. (Eds.). (1969). *Symposium on theory and applications of generalized inverses of matrices, Texas Technological College, Lubbock, 1968. Proceedings.* Lubbock: Texas Technological Press.

Boullion, T. L., & Odell, P. L. (1971). *Generalized inverse matrices.* New York: Wiley-Interscience.

Boyer, C. B. (1956). *History of analytic geometry (scripta mathematica).* New York: Yeshiva University.

Brand, L. (1962). The solution of linear algebraic equations. *Mathematical Gazette, 46,* 203–207.

Breneman, E. J., & Bartleson, C. J. (1966). "Luminosity" and "brightness." *Journal of the Optical Society of America, 56,* 983.

Brewer, W. L., & Holly, F. R., Jr. (1948). Synthesis of spectral distribution curves. *Journal of the Optical Society of America, 38,* 858–874.

Brewster, D. (1828). Account of the performances of different ventriloquists with observations on the art of ventriloquism. *Edinburgh Journal of Science, 9,* 252–259.

Brewster, D. (1831). On a new analysis of solar light, indicating the three primary colours, forming coincident spectra of equal length. *Edinburgh Journal of Science, n.s., 5*, 197–206.

Brill, M. H. (1985). Decomposition of Cohen's matrix **R** into simpler color invariants. *American Journal of Psychology, 98*, 625–634.

Brill, M. H. (1987). Statistical confirmation of Thornton's zero-crossing conjecture for metameric blacks. *Color Research and Application, 12*, 51–52.

Brill, M. H. (1996). Do tristimulus values have units? *Color Research and Application, 21*, 310–313.

Brill, M. H., & West, G. (1986). Chromatic adaptation and color constancy: A possible dichotomy. *Color Research and Application, 11*, 196–204.

Bronowski, J. (1973). *The ascent of man*. Boston: Little, Brown.

Burch, R. M. (1910). *Colour printing and colour printers*. London: Pitman.

Burnham, R. W., Hanes, R. M., & Bartleson, C. J. (1963). *Color: A guide to basic facts and concepts*. New York: Wiley.

Burns, S. A., Cohen, J. B., & Kuznetsov, E. N. (1989). Multiple metamers: Preserving color matches under diverse illuminants. *Color Research and Application, 14*, 16–22.

Burns, S. A., Cohen, J. B., & Kuznetsov, E. N. (1990). The Munsell color system in fundamental color space. *Color Research and Application, 15*, 25–51.

Burns, S. A., Smith, V. C., Pokorny, J., & Elsner, A. E. (1982). Brightness of equal-luminance lights, *Journal of the Optical Society of America, 72*, 1225–1231.

Campbell, L., & Garnett, W. (1882). *The life of James Clerk Maxwell*. London: Macmillan.

Campbell, S. L., & Meyer, C. D., Jr. (1978). *Generalized inverses of linear transformations*. London: Pitman.

Castel, L. B. [attributed]. (1723, June). Review of *Traité d'optique* by Isaac Newton. *Journal de Trévoux*, 1428–1450.

Castel, L. B. (1735, various). Nouvelles expériences d'optique et d'acoustique. *Journal de Trévoux*, August, 1444–1482, 1619–1666; September, 1807–1839; October, 2018–2053; November, 2333–2372; December, 2642–2768.

Castel, L. B. (1737, August). Review of *Coloritto or the Harmony of Colouring* by Le Blon. *Journal de Trévoux*, 1435.

Castel, L. B. (1740). *L'Optique des couleurs, fondée sur les simples observations, et tournée surtout à la pratique de la peinture, de la teinture et des autres arts coloristes*. Paris: Briasson. (German translation. *Die auf langer Erfahrung gegründete Farben-Optik*. Halle, 1747.)

Castel, L. B. (1743). *Le Vrai système de physique générale de M. Isaac Newton exposé et analysé en parallèle avec celui de Descartes*. Paris: C. F. Simon fils.

Chapanis, A., & Halsey, R. M. (1955). Luminance of equally bright colors. *Journal of the Optical Society of America, 45*, 1–6.

Chevreul, M. E. (1839). *De la loi du contraste simultané des couleurs et de l'assortiment des objets colorés considéré d'après cette loi*. Paris: Pitois-Levrault. (Many translations including M. E. Chevreul, *The Laws of Contrast of Colour and Their Application to the Arts*. London: Routledge, Warnes, and Routledge, 1861 and M. E. Chevreul, *The Principles of Harmony and Contrast of Colors and their Applications to the Arts*. NewYork: Reinhold, 1967.)

CIE. *See* Commission Internationale de l'Éclairage.

Coblentz, W. W., & Emerson, W. B. (1918–1919). Relative sensibility of the average eye to light of different colors and some practical applications to radiation problems. *Bulletin of the Bureau of Standards (United States), 14*, 167–236.

Cohen, I. B. (1958). *Isaac Newton's papers and letters on natural philosophy*. Cambridge, MA: Harvard University Press.

Cohen, J. B. (1988). Color and color mixture: Scalar and vector fundamentals. *Color Research and Application, 13*, 5–39.

Cohen, J. B., & Kappauf, W. E. (1982). Metameric color stimuli, fundamental metamers, and Wyszecki's metameric blacks. *American Journal of Psychology, 95*, 537–564.

Cohen, J. B., & Kappauf, W. E. (1985). Color mixture and fundamental metamers: Theory, algebra, geometry, application. *American Journal of Psychology, 98*, 171–259.

Cohen, J. B., & Walker, R. J. (1946). The transformation matrix of color mixture data. *American Journal of Psychology, 59*, 299–305.

Cohen, M. R., & Drabkin, I. E. (1948). *A source book in Greek science*. Cambridge, MA: Harvard University Press.

Commission Internationale de l'Éclairage. (1983). *The basis of physical photometry* (2nd ed.). Vienna: Central Bureau of the CIE.

Commission Internationale de l'Éclairage. (1986). *Colorimetry* (2nd ed.). Vienna: Central Bureau of the CIE.

Commission Internationale de l'Éclairage. (1987). *International lighting vocabulary* (4th ed.). Vienna: Central Bureau of the CIE.

Cornell, E. S. (1938a). The radiant heat spectrum from Herschel to Melloni, I: The work of Herschel and his contemporaries. *Annals of Science, 3*, 119–137.

Cornell, E. S. (1938b). The radiant heat spectrum from Herschel to Melloni, II: The work of Melloni and his contemporaries, *Annals of Science, 3*, 402–416.

Crew, H. (1930). Thomas Young's place in the history of the wave theory of light. *Journal of the Optical Society of America, 20*, 3–10.

Crittenden, E. C., & Richtmeyer, F. K. (1918–1919). An "average eye" for heterochromatic photometry, and a comparison of a flicker and an equality-of-brightness photometer. *Bulletin of the Bureau of Standards (United States), 14*, 87–113.

Crowe, M. J. (1967). *A history of vector analysis: The evolution of the idea of a vectorial system*. Notre Dame, IN: University of Notre Dame Press.

Desaguliers, J. T. (1716). An account of light and colours, formerly made by Sir Isaac Newton, and mentioned in his Opticks, lately repeated before the Royal Society. *Philosophical Transactions of the Royal Society of London, 29*, 433–447.

Desaguliers, J. T. (1722). An account of an optical experiment made before the Royal Society, on Thursday, Dec. 6th and repeated on the 13th, 1722. *Philosophical Transactions of the Royal Society of London, 32*, 206–208.

Desaguliers, J. T. (1734–1744). *A course of experimental philosophy*. (2 vols.). London: Vol. 1 printed for J. Senex; Vol. 2 printed for W. Innys; 2d ed., corr. London: W. Innys, 1745, 2 vols.; 3d ed., corr. London, printed for A. Millar, 1763, 2 vols.

Desmaizeaux, P. (1722). Lettre écrite de Londres, touchant l'art d'imprimer des tableaux et des portraits en couleurs. *Journal des Sçavans, 72*, 46–49.

Diels, H. (1934–1937). *Die Fragmente der Vorsokratiker*. (W. Kranz, Ed.). (5th ed., 3 vols.) Berlin: Weidmannsche Buchhandlung.

Dobrowolsky, W. (1872). Über die Empfindlichkeit des Auges gegen die Lichtintensität verschiedener Spectralfarben. *Archiv. Ophthalmol., 18*(1) 74–92.

Dove, H. W. (1852). Über den Einfluss der Helligkeit einer weissen Beleuchtung auf die relative Intensität verschiedener Farben. *Annalen der physikalischen Chemie, 161*, 397–408.

Dresler, A. (1953). The non-additivity of heterochromatic brightness. *Transactions of the Illuminating Engineering Society (London), 18*, 141–165.

Dziobek, W. (1926). Über die Transformation von einem trichromatischen System auf ein anderes. *Zeitschrift für Instrumentenkunde, 46*, 81–84.

Engel, F. (1911). Grassmann's Leben. In H. G. Grassmann, *Gesammelte Mathematische und Physikalische Werke* (Vol. 3, pt. 2). Leipzig: B. G. Teubner.

Estévez, O. (1982). A better colorimetric standard observer for color-vision studies: The Stiles and Burch 2° color matching functions. *Color Research and Application, 7*, 131–134.

Evans, R. M. (1948). *An introduction to color.* New York: Wiley.

Evans, R. M. (1974). *The perception of color.* New York: Wiley-Interscience.

Everitt, C. W. F. (1975). *James Clerk Maxwell, physicist and natural philosopher.* New York: Scribner's.

Fairman, H. S. (1986). Matrix-R conference. *Color Research and Application, 11*, 295.

Fairman, H. S. (1987). Metameric correction using parametric decomposition. *Color Research and Application, 12*, 261–265.

Feynman, H. R. P., Leighton, R. B., & Sands, M. (1963). *The Feynman lectures on physics.* Reading, MA: Addison-Wesley.

Ficken, F. A. (1967). *Linear transformations and matrices.* Englewood Cliffs, NJ: Prentice-Hall.

Field, G. (1835). *Chromatography; or A treatise on colours and pigments, and of their powers in painting, etc.* London: C. Tilt.

Field, G. (1845). *Chromatics; or The analogy, harmony, and philosophy of colours.* London: David Bogue.

Fox, A. H. (1963). *Fundamentals of numerical analysis.* New York: Ronald.

Fraunhofer, J. (1814–1815). Bestimmung des Brechungs-und Farbenzerstreungs-Vermögens verschiedner Glassarten, in bezug auf die achromatische Fernröhre. *Denkschriften der bayrischen Akademie der Wissenschaften, München, Mathematisch-Naturwissenschaftliche Classe, 5*, 193–226.

Fraunhofer, J. von. (1888). *Gesammelte Schriften.* Munich: Verlag der Königlich Bayrischen Akademie der Wissenschaften.

Fraunhofer, J. von. (1898). *Prismatic and diffraction spectra: Memoirs.* (J. S. Ames, Trans.) New York: Harper & Brothers.

Freeman, K. (1966). *Ancilla to the pre-Socratic philosophers. A complete translation of the fragments in Diels' Fragmente der Vorsokratiker.* Cambridge, MA: Harvard University Press.

Frey, M. v., & Kries, J. v. (1881). Ueber die Mischung von Spectralfarben. *Archiv für Anatomie und Physiologie*, Physiologische Abtheilung, 336–353.

Froelich, C. E. (1924). Algebraic methods for the calculation of color mixture transformation diagrams. *Journal of the Optical Society of America, 9*, 31–42.

Gautier d'Agoty, J. F. (1749a). *Chroa-génésie ou génération des couleurs, contre le systeme de Newton.* Paris: Delaguette.

Gautier d'Agoty, J. F. (1749b). *Lettres concernant le nouvel art de graver et d'imprimer les tableaux.* Paris: J. Bullot.

Gautier d'Agoty, J. F. (1752). *Rèfutation de la défense des newtoniens.* Paris: Delaguette.

Gautier d'Agoty, J. F. (1752–1755). *Obervations sur l'histoire naturelle, sur la physique et sur la peinture.* Paris: Delaguette.

Genberg, E. (1933). Some complementary colour relations. *Proceedings of the Physical Society of London, 45*, 836–840.

Gibson, G. L. (1981). Color-matching functions: A set of transformations with essentially single maximums. *Color Research and Application, 5*, 143–146.

Gibson, K. S. (1948). Spectral luminosity factors. *Journal of the Optical Society of America, 38*, 51–61.

Gibson, K. S., & Tyndall, E. P. T. (1923–1924). Visibility of radiant energy. *Scientific Papers of the Bureau of Standards (United States), 19*, 131–191.

Gladstone, W. E. (1858). *Studies on Homer and the Homeric age* (vol. 3). Oxford: Oxford University Press.

Glencross, M. J. (1979). Calculating generalized inverse matrices. *Mathematical Gazette, 63*, 173–175.

Godlove, I. H. (1930). The wavelengths of complementary hues. *Journal of the Optical Society of America, 20*, 411–418.

Goldhammer, D. A. (1905). Die Farbenempfindlichkeit des Auges und die photometrische Helligkeit der leuchtenden Körper. *Annalen der Physik, 16*, 621–652.

Graham, C. H. (1965). Color mixture and color systems. In C. H. Graham (Ed.), *Vision and visual perception* (pp. 370–394). New York: Wiley.

Grailich, J. (1854). Beitrag zur Theorie der gemischten Farben. *Sitzungsberichte der Akademie der Wissenschaften, Wien, Mathematisch-Naturwissenschaftliche Classe, 13*, 201–284.,

Gram, J. P. (1883). Über die Entwicklung reeller Functionen in Reihen mittelst der Methode der kleinsten Quadrate. *Journal für die reine und angewandte Mathematik, 14* (94), 41–73.

Grassmann, H. G. (1847). *Die Geometrische Analyse gekúpft und die von Leibnitz erfundene Geometrische Charakteristik.* Leipzig: Jablonowskische Gesellschaft der Wissenschaft.

Grassmann, H. G. (1853). Zur Theorie der Farbenmischung, *Poggendorf's Annalen der Physik und Chemie* (series 3), *29*, 69–84. (Translation: On the theory of compound colours, *Philosophical Journal Magazine*, 1854 (series 4), *7*, 254–264.)

Grassmann, H. G. (1877). Bemerkungen zur Theorie der Farbenempfindungen. In W. Preyer, *Elemente der reinen Empfindungslehre* (pp. 86–93). Jena, Germany: Herman Dufft.

Grassmann, H. G. (1894–1911). *Gesammelte Mathematische und Physikalische Werke.* (F. Engel, Ed.). (3 vols. in 6 pts.). Leipzig: B. G. Teubner.

Grünberg, V. (1904). Farbengleichung mit Zuhilfenahme der drei Grundempfindungen im Young-Helmholtzschen Farbensystem. *Sitzungberichte der Kaiserlichen Akademie der Wissenschaften*, Wien, II A, *113*, 627–636.

Guild, J. (1924–1925a). The transformation of trichromatic mixture data: Algebraic methods. *Transactions of the Optical Society (London), 26*, 95–108.

Guild, J. (1924–1925b). The geometrical solution of colour mixture problems. *Transactions of the Optical Society (London), 26*, 139–174.

Guyot, E. G. (1769–1770). *Nouvelles récréations physiques et mathématiques.* (4 vols.). Paris: Gueffier.

Hamilton, W. R. (1853). *Lectures on quaternions.* Dublin: Hodges and Smith.

Hanson, W. T., & Brewer, W. L. (1954a). Subtractive color photography: The role of masks. *Journal of the Optical Society of America, 44,* 129–134.

Hanson, W. T., & Brewer, W. L. (1954b). Subtractive color photography: Spectral sensitivities and masks. *Journal of the Optical Society of America, 44,* 476–481.

Hardy, A. C., & MacAdam, D. L. (1936). *Handbook of colorimetry.* Cambridge, MA: MIT Press.

Hardy, A. C., & Wurzburg, F. L. (1937). The theory of three-color reproduction. *Journal of the Optical Society of America, 27,* 227–240.

Hecht, S. (1924). The visibility of the spectrum. *Journal of the Optical Society of America, 9,* 211–222.

Hecht, S. (1930). The development of Thomas Young's theory of color vision. *Journal of the Optical Society of America, 20,* 231–270.

Hellmig, E. (1954). Die bedingt-gleichen Farben und die Farben mit Farbwerten Null. *Die Farbe, 3,* 165–174.

Hellmig, E. (1955). Praktische Methoden zur Ermittlung der Remissionsfunktionen von Farben mit vorgegebenen Farbwerten, unter besonderer Berücksichtigung von "Unecht Schwarz." *Die Farbe, 4,* 3–14.

Helmholtz, H. von (1852). Über die Theorie der zusammengesetzten Farben. *J. C. Poggendorff's Annalen der Physik und Chemie, 87,* 45–66. (English translation: On the theory of compound colours. *Philosophical Magazine, 4,* 519–534 [1852].)

Helmholtz, H. von (1855). Über die Zusammensetzung von Spectralfarben. *J. C. Poggendorff's Annalen der Physik und Chemie, 94,* 1–28.

Helmholtz, H. von (1856–1867). *Handbuch der Physiologischen Optik.* Leipzig: L. Voss. 2nd ed. Hamburg and Leipzig: L. Voss, 1896. Translated and edited from the 3rd German ed. by J. P. C. Southall: *Helmholtz's Treatise on Physiological Optics.* Rochester, N.Y.: The Optical Society of America, (1924–25).

Helmholtz, H. von (1873). The recent progress in the theory of vision. In H. v. Helmholtz, *Popular Lectures on Scientific Subjects.* (B. A. Pye-Smith & E. Atkinson, Trans.). New York: Appleton.

Hering, E. (1887). Über Newton's Gesetz der Farbenmischung. *Jahrbuch "Lotos," 7,* Prague: Tempsky.

Herschel, W. (1800a). Investigation of the powers of the prismatic colours to heat and illuminate objects; with remarks that prove the different refrangibility of radiant heat, to which is added, an inquiry into the method of viewing the sun advantageously with telescopes of large apertures and high magnifying powers. *Philosophical Transactions of the Royal Society of London, 90,* 255–283.

Herschel, W. (1800b). Experiments on the refrangibility of the invisible rays of the sun. *Philosophical Transactions of the Royal Society of London, 90,* 284–292.

Herschel, W. (1800c). Part I. Experiments on the solar, and on the terrestrial rays that occasion heat; with a comparative view of the laws to which light and heat, or rather the rays which occasion them, are subject, in order to determine whether they are the same or different. *Philosophical Transactions of the Royal Society of London, 90,* 293–326, 437–538.

Herschel, J. W. F. (1845). *Light.* In *Encyclopaedia Metropolitania.* (Vol. 4, 341–586). London. [Herschel himself referred to this as his "Treatise on Light." 1st ed., 1827; later ed., 1849.]

Hertz, H. R. (1894–1895). *Gesammelte Werke.* (3 vols.). Leipzig: J. A. Barth. (Translated as *Miscellaneous Papers,* by D. E. Jones and G. A. Schott. New York: Macmillan, 1896.)

Hickerson, N. P. (1971). Review of basic color terms. *International Journal of American Linguistics, 37,* 257–270.

Holton, G. (1952). *Introduction to concepts and theories in physical science.* Reading, MA: Addison-Wesley.

Houston, R. A. (1930). The visibility of radiation and dark adaptation. *Philosophical Magazine,* 7th ser., *10,* 416–432.

Houston, R. A. (1931). Color vision. *Philosophical Magazine,* 7th ser., *11,* 552–569.

Houston, R. A. (1932). *Vision and colour vision.* London: Longman, Green.

Howett, G. L., & Graham, C. H. (1957). *Transformations of trichromatic coordinates in colorimetry. Research report.* New York: Psychological Laboratory, Columbia University.

Hübl, A. F. (1897). *Die Dreifarbenphotographie, mit besonderer Berücksichtigung des Dreifarbendruckes und der photographischen Pigmentbilder in natürlichen Farben.* Halle a.s.: W. Knapp (Translation: H. O. Klein. *Three-Colour Photography.* London: Penrose, 1904).

Hulbert, E. O. (1926). The transparency of ocean water and the visibility curve of the eye. *Journal of the Optical Society of America and Review of Scientific Instruments, 13,* 553–556.

Hurvich, L. M., & Jameson, D. (1953a). Spectral sensitivity of the fovea. I. Neutral adaptation. *Journal of the Optical Society of America, 43,* 485–494.

Hurvich, L. M., & Jameson, D. (1953b). Spectral sensitivity of the fovea. II. Dependence on chromatic adaptation. *Journal of the Optical Society of America, 43,* 552–559.

Hurvich, L. M., & Jameson, D. (1954). Spectral sensitivity of the fovea. III. Heterochromatic brightness and chromatic adaptation. *Journal of the Optical Society of America, 44,* 213–222.

Huygens, C. (1673). An extract from a letter written by an ingenious person from Paris, containing some Considerations upon Mr. Newton's doctrine of colors [*sic*], as also upon the effects of the different refractions of the rays in telescopical glasses. *Proceedings of the Royal Society (London), 96,* 6086–6087.

Hyde, E. P., & Forsythe, W. E. (1915). The visibility of radiation in the red end of the visible spectrum. *Astrophysical Journal, 42,* 285–293.

Hyde, E. P., Forsythe, W. E., & Cady, F. E. (1918). The visibility of radiation. *Astrophysical Journal, 48,* 65–88.

Imai, F. H., & Berns, R. S. (1998). High-resolution multi-spectral image archives: A hybrid approach. *Proceedings of the Sixth IS&T/SID Color Imaging Conference,* Scottsdale, AZ.

Ives, J. E. (1900). The optics of trichromatic photography. *Photographic Journal, 40,* 99–121.

Ives, H. E. (1912a, July). Studies in the photometry of lights of different colors. I: Spectral luminosity curves obtained by the equality of brightness photometer and the flicker photometer under similar conditions. *Philosophical Magazine,* 6th ser., *24,* 149–187.

Ives, H. E. (1912b, September). Studies in the photometry of lights of different colors. II: Spectral luminosity curves obtained by the method of critical frequency. *Philosophical Magazine,* 6th ser., *24,* 352–370.

Ives, H. E. (1912c, November). Studies in the photometry of lights of different colors. III: Distortions in spectral luminosity curves produced by variations in the character of the comparison standard and the surroundings of the photometric field. *Philosophical Magazine*, 6th ser., *24*, 744–751.

Ives, H. E. (1912d, December). Studies in the photometry of lights of different colors. IV: The addition of luminosities of different color. *Philosophical Magazine*, 6th ser., *24*, 845–853.

Ives, H. E. (1912e, December). Studies in the photometry of lights of different colors. V. The spectral luminosity curve of the average eye. *Philosophical Magazine*, 6th ser., *24*, 853–863.

Ives, H. E. (1915, December). The transformation of color mixture equations from one system to another. *Journal of the Franklin Institute, 180*, 673–701.

Ives, H. E. (1923). The transformation of color mixture equations from one system to another. II. Graphical aids. *Journal of the Franklin Institute, 195*, 23–44.

Ives, H. E., & Kingsbury, E. F. (1914, November). The theory of the flicker photometer. *Philosophical Magazine*, 6th ser., *28*, 708–729.

Ives, H. E., & Kingsbury, E. F. (1916, April). The theory of the flicker photometer. II: Unsymmetrical conditions. *Philosophical Magazine*, 6th ser., *31*, 290–321.

James, M. (1978). The generalised inverse. *Mathematical Gazette, 62*, 109–116.

Judd, D. B. (1924). The spectral energy distributions produced by rotary mixing of complementary papers. *Journal of the Optical Society of America and Review of Scientific Instruments, 9*, 95–111.

Judd, D. B. (1930). Reduction of data on mixture of color stimuli. *Journal of Research, United States Bureau of Standards* (Research Paper 163), *4*, 515–548.

Judd, D. B. (1931). Extension of the standard visibility function to intervals of 1 millimicron by third-difference osculatory interpolation. *U.S. Bureau of Standards, Journal of Research, 6*, 465–471.

Judd, D. B. (1951). Report of U.S. Secretariat. Committee on Colorimetry and Artificial Daylight. *CIE Proceedings, 12th Session, Stockholm*. Paris: Bureau Central de la CIE.

Judd, D. B. (1964). Maxwell and modern colorimetry. *Journal of Photographic Science, 9*, 341–352.

Judd, D. B., & Kelly, K. L. (1939). Method of designating colors. *Journal of Research, National Bureau of Standards, 23*, 355–366.

Kaiser, P. K. (1971). Luminance and brightness. *Applied Optics, 10*, 2768–2770.

Kaiser, P. K., & Wyszecki, G. (1978). Additivity failures in heterochromatic brightness matching. *Color Research and Application, 3*, 177–182.

Kay, P., & McDaniel, C. K. (1978). The linguistic significance of the meanings of basic color terms. *Language, 54*, 610–646.

Kingsbury, E. F. (1916). A spectral luminosity curve equation and its use. *Physical Review*, 2nd ser., *7*, 161–168.

Kirschmann, A. (1917). Das umgekehrte Spectrum und seine Komplementärverhältnisse, *Physikalische Zeitschrift, 18*, 195–205.

König, A. P. (1891). Über den Helligkeitswert der Spektralfarben bei verschiedener absoluter Intensität. *Beiträge zur Psychologie und Physiologie der Sinnesorgane*, 309–388.

König, A. P. (1894). Über die bisher gemachten Bestimmungen der Wellenlängen einfacher complementärer Farben. *Verhandlungen der physikalischen Gesellschaft, Berlin, 9*, 37–39.

König, A. P. (1896). Quantitative Bestimmungen an complementären Spectralfarben. *Sitzungsberichte der Akademie der Wissenschaften zu Berlin, 30*, 945–949.

König, A. P., & Dieterici, C. (1886, July 29) Die Grundempfindungen und ihre Intensitätsvertheilung im Spectrum. *Sitzungsberichte der Akademie der Wissenschaften zu Berlin*, 805–829.

Kuehni, R. G. (1979). Intersection nodes of metameric matches. *Color Research and Application, 4*, 101–102.

Kuehni, R. G., & Berns, R. S. (1994). The determinants of metameric crossovers, revisited. *Color Research and Application, 19*, 392–394.

Lambert, J. H. (1760). *Photometria*. Augsburg.

Lang, H. (1980). Die Farbensysteme des 18. Jahrhunderts von Mayer, Lambert, und Lichtenberg. *Farbe und Design, 15/16*, 50–59.

Lang, H. (1983). Trichromatic theories before Young. *Color Research and Application, 8*, 221–231.

Langley, S. P. (1889). Energy and vision. *Philosophical Magazine*, 5th ser., *27–28*, 1–23.

Le Blon, J. C. (1756). *Coloritto, or the harmony of colouring in painting, reduced to mechanical practice, under easy precepts, and infallible rules.* Combined English and French ed. London, circa 1725. Reprinted in English and French, with introd. by Faber Birren, New York: Van Nostrand, Reinhold, 1980.

Le Grand, Y. (1957). *Light, colour, and vision.* (R. W. G. Hunt, J. W. T. Walsh, & F. R. W. Hunt, Trans.). New York: Wiley.

Le Grand, Y. (1972). Spectral luminosity. In D. Jameson & L. M. Hurvich (Eds.), *Visual psychophysics* (pp. 413–433). Berlin: Springer-Verlag.

Leicester, H. M. (1970). *Mikhail Vasil'evich Lomonosov on the corpuscular theory.* Cambridge, MA: Harvard University Press.

Lewis, A. C. (1977). Grassmann's 1844 *Ausdehnungslehre* and Schleiermacher's *Dialektik. Annals of Science, 34*, 103–162.

Lomonosov, M. V. (1747). *Discourse on the generation of light, representing a new color theory.* Delivered in Russian to the Moscow Academy in 1747. English translation, Leicester, 1970, above.

Longley, J. W. (1984). *Least squares computations using orthogonalization methods.* New York: Marcel Dekker.

MacAdam, D. L. (1937). Projective transformations of I.C.I. color specifications. *Journal of the Optical Society of America, 27*, 294–299.

MacAdam, D. L. (1938). Photometric relationships between complementary colors. *Journal of the Optical Society of America, 28*, 103–111.

MacAdam, D. L. (1942). Projective transformations of color-mixture diagrams. *Journal of the Optical Society of America, 32*, 2–6.

MacAdam, D. L. (1950). Loci of constant hue and brightness determined with various surrounding colors. *Journal of the Optical Society of America, 40*, 589–595.

MacAdam, D. L. (1951). Influence of visual adaptation on loci of constant hue and saturation. *Journal of the Optical Society of America, 41*, 615–619.

MacAdam, D. L. (1953). Dependence of color-mixture functions on choice of primaries. *Journal of the Optical Society of America, 43*, 533–538.

MacAdam, D. L. (1954). Orthogonal color-mixture functions. *Journal of the Optical Society of America, 44*, 713–724.

MacAdam, D. L. (1975). Color essays. *Journal of the Optical Society of America, 65*, 483–492.

MacAdam, D. L. (1983). Luminosity as a color-matching function. *Color Research and Application, 8*, 121.

MacAdam, D. L., & Shapiro, A. E. (1985). Newton's theory of color. *Physics Today, 38*, 11–15.

Marat, J. P. (1780). *Découvertes sur la lumière*. London and Paris: Jombert.

Marat, J. P. (1784). *Notions élémentaires d'optique*. Paris: P. F. Didot, N. L. Moutard.

Mariotte, E. (1717). De la nature des couleurs. In *Oeuvres de M. Mariotte*. Leiden: P. Van der Aa.

Maxwell, J. C. (1854–1855). Experiments on colour, as perceived by the eye, with remarks on colour blindness. *Transactions of the Royal Society of Edinburgh, 21 pt. 2*, 275–298.

Maxwell, J. C. (1860). On the theory of compound colours, and the relations of the colours of the spectrum. *Philosophical Transactions of the Royal Society of London, 150B*, 57–84.

Maxwell, J. C. (1871). On colour vision. *Nature (London), 4*, 13–16.

Maxwell, J. C. (1872). On colour vision. *Proceedings of the Royal Society of Great Britain, 6*, 260–271.

Mayer, A. M. (1875). The history of Young's discovery of his theory of colours. *American Journal of Science and Arts*, ser. 3, *9*, 251–267.

Miller, C. W. (1941). The matrix algebra and color reproduction. *Journal of the Optical Society of America, 31*, 477–482.

Millington, E. C. (1941). History of the Young-Helmholtz theory of vision. *Annals of Science, 5*, 167–177.

Möbius, A. F. (1827). *Der Barycentrische Calcul*. In *August Ferdinand Möbius Gesammelte Werke*, edited by R. Baltzer. Leipzig: Verlag von Johann Ambrosius Barth.

Montdorge, G. de (1751–1765). Gravure en couleurs, à l'imitation de la peinture. In D. Diderot & J. L. D'Alembert, *Encyclopédie, ou Dictionnaire Raisonné, des Sciences, des Arts, et des Métiers*. Paris: Briasson, David, Lebreton, Durand.

Moon, P., & Spencer, D. E. (1933). Analytical representation of standard response curves. *Journal of the Optical Society of America, 33*, 89–103.

Moore, E. H. (1920). The algebra of matrices, Part I of the theory of general analysis. *Bulletin of the American Mathematical Society, 26*, 394–395.

Morris, R. H. (1947). Metameric formulation. *Journal of the Optical Society of America, 37*, 669.

Mortimer, C. (1731). An account of Mr. James Christopher Le Blon's principles of printing in imitation of painting and of weaving tapestry, in the same manner as brocades. *Philosophical Transactions of the Royal Society (London), 419*, 101–107.

Müller, J. (1826). *Zur vergleichenden Physiologie des Gesichtssinnes des Menschen und der Thiere*. Leipzig: C. Cnobloch.

Nashed, M. Z. (Ed.). (1976). *Generalized inverses and their applications*. New York: Academic Press.

Neugebauer, H. E. J. (1956). Quality factor for filters whose spectral transmittances are different from color mixture curves, and its application to color photography. *Journal of the Optical Society of America, 46*, 821–824.

Newton, I. (1671–1672). New theory about light and colors [*sic*]. *Philosophical Transactions of the Royal Society (London), 6*, 3075–3087.

Newton, I. (1673). An extract of Mr. Isaac Newton's Letter, written to the publisher from Cambridge April 3, 1673, concerning the number of colors [*sic*], and the necessity of mixing them all for the production of white . . . which letter, being an immediate answer to that from Paris . . . should also, if it had not been mislaid, have immediately followed the same. *Transactions of the Royal Society (London), 77*, 6108–6111.

Newton, I. (1675–1676). Newton's second paper on color and light, read at the Royal Society in 1675/76. In T. Birch, *The history of the Royal Society of London* (Vol. 3). London: A. Millar, 1757.

Newton, I. (1704). *Opticks: or, a treatise of the reflexions, refractions, inflexions and colours of light.* London: Sam Smith and Benj. Walford, Printers to the Royal Society.

Niven, W. D. (1879). On certain definite integrals occuring in spherical harmonic analysis, and on the expansion in series of the potentials of the ellipsoid and of the ellipsis. *Proceedings of the Royal Society, 29*, 2–6.

Nutting, P. G. (1908–1909). The luminous equivalent of radiation. *Bulletin of the Bureau of Standards (United States), 5*, 261–308.

Nutting, P. G. (1911). The visibility of radiation: A recalculation of König's data. *Bulletin of the Bureau of Standards (United States) 7*, 235–238.

Nutting, P. G. (1915). The visibility of radiation. *Philosophical Magazine*, 6th ser., *29–30*, 301–309.

Ohta, N. (1982). Practical transformations of CIE color-matching functions. *Color Research and Application, 7*, 53–56.

Ohta, N., & Wyszecki, G. (1977). Location of the nodes of metameric color stimuli. *Color Research and Application, 2*, 183–186.

Ohyama, N., Yachida, M., Badique, E., Tsujiuchi, J., & Honda, T. (1988). Least-squares filter for color-image restoration. *Journal of the Optical Society of America, 5A*, 19–24.

Onley, J. W., & Sternheim, C. E. (1967). Psychophysical responses to homochromatic stimuli of equal brightness but unequal luminance. *Journal of the Optical Society of America, 57*, 258–267.

Ostwald, W. (1919). *Die Farbenlehre* and *Physikalische Farbenlehre*, Leipzig: Verlag Unesma. (Both books were published in one cover.)

Palmer, G. (1777). *Theory of colours and vision.* London: S. Leacroft.

Palmer, G. (1786). *Théorie de la Lumière, applicable aux Arts, et principalement à la Peinture.* Paris: Hardouin et Gattey.

Parsons, J. (1930–1931). Young's theory of colour vision (1801–1931). *Transactions of the Optical Society of London, 32*, 165–183.

Peacock, G. (1855). *Life of Thomas Young, M.D., F.R.S. &c.*, London: John Murray.

Pearson, M. L., & Yule, J. A. C. (1973). Transformation of color mixture functions without negative portions. *Journal of Color and Appearance, 2*, 30–35.

Peddie, R. A. (1917). *An outline of the history of painting to which is added the history of printing in colours.* London: Grafton.

Peddie, W. (1922). *Color vision.* London: Edward Arnold.

Penrose, R. (1956). A generalized inverse for matrices. *Proceedings, Cambridge Philosophical Society, 51*, 406–413.

Pflüger, A. (1902). Über die Farbenempfindlichkeit des Auges. *Annalen der Physik*, 4th series, *9*, 185–208.

Planitz, M. (1979). Inconsistent systems of linear equations. *Mathematical Gazette, 63*, 181–185.

Pratt, W. K. (1978). *Digital imaging processing.* New York: Wiley.

Preyer, W. (1877). *Elemente der reinen Empfindungslehre.* Jena, Germany: Hermann Dufft.

Pridmore, R. W (1978). Complementary colors: Composition and efficiency in producing various whites. *Journal of the Optical Society of America, 68*, 1490–1496.

Pridmore, R. W. (1980). Complementary colors: A correction. *Journal of the Optical Society of America, 70*, 248–249.

Pridmore, R. W. (1988). Complementary colors: Correction: errata. *Journal of the Optical Society of America, A5*, 884.

Priest, I. G. (1921–1922). Note on the relations between the frequencies of complementary hues. *Journal of the Optical Society of America, 4*, 402–404, 5, 513.

Priestley, J. (1772). *The history and present state of discoveries relating to vision, light, and colours.* London: J. Johnson.

Purkinje, J. E. (1825). *Beobachtungen und Versuche zur Physiologie der Sinne. Vol. 2: Neue Beiträge zur Kenntniss des Sehens in subjektiver Hinsicht.* Berlin: Reimer.

Rao, C. R., & Mitra, S. K. (1971). *Generalized inverse of matrices and its applications.* New York: Wiley.

Reade, J. (1814). Experiments tending to prove, that neither Sir Isaac Newton, Herschel, nor any other person, ever decomposed incident or impingent light into the prismatic colours. *Philosophical Magazine, 43*, 193–197.

Richter, M. (1958). Remissionskurven bedingt-gleicher Farben. *Die Farbe, 7*, 163–165.

Rood, O. N. (1878). On the photometric comparison of lights of different colors. *American Journal of Science and Arts, 65*, 81–82.

Rood, O. N. (1895). *Student's text-book of color; or, modern chromatics with applications to art and industry.* New York: D. Appleton.

Ross, R. J. (1970). *Color film for color television.* New York: Hastings House.

Runge, I. (1928). Über die Ermittlung der Farbkoordinaten aus den Messungen am trichromatischen Kolorimeter. *Zeitschrift für Instrumentenkunde, 48*, 387–396.

Sanders, C. L., & Wyszecki, G. (1957). Correlate for lightness in terms of CIE-tristimulus values. Part I. *Journal of the Optical Society of America, 47*, 398–404.

Sawyer, W. W. (1981). Colour perception and a metric. *Mathematical Gazette, 65*, 54–56.

Schier, D. S. (1941). *Louis Bertrand Castel, anti-Newtonian scientist.* Cedar Rapids, IA: Torch Press.

Schlegel, V. (1878). *Hermann Grassmann, sein Leben und seine Werke.* Leipzig: Brockhaus.

Schmidt, E. (1907). Zur Theorie der linearen und nichtlinearen Integralgleichungen. I. Teil: Entwicklung willkütlicher Functionen nach Systemen vorgeschreibenerart. *Mathematische Annalen, 63*, 433–476.

Schmitt, F. J. M. (1976). A method for the treatment of metamerism in colorimetry. *Journal of the Optical Society of America, 66*, 601–608.

Schmitt, F. J. M. (1977a) A statistical generation of metamers (pp. 436–438). In *Color 77 (3rd A.I.C, Congress)*, Troy, NY.

Schmitt, F. J. M. (1977b) *Contribution à l'étude du metamérisme.* Unpublished doctoral dissertation, Université Pierre et Marie Curie.

Schopenhauer, A. (1816). *Über das Sehen und die Farben.* Leipzig: J. F. Hartknock.

Schrödinger, E. (1920a). Theorie der Pigmente von grösster Leuchtkraft. *Annalen der Physik, 62*, 603–622.

Schrödinger, E. (1920b). Grundlinien einer Theorie der Farbenmetrik im Tagessehen. *Annalen der Physik, 63*, 397–426, 427–456, 481–520.

Schrödinger, E. (1925). Über das Verhältnis der Vierfarben-zur Dreifarbentheorie, *Sitzungsberichte der Akademie der Wissenschaften, Wien*, Mathematische-Naturwissenschaftliche Klasse, Abteilung 2a *134*, 417.

Seebeck, A. (1837). Über den bei manchen Personen vorkommenden Mangel an Farbensinn. *Annalen der Physik und Chemie*, 2nd series, *118*, 177–233.

Shapiro, A. E. (1980). The evolving structure of Newton's theory of white light and color. *Isis, 71*, 211–235.

Shapiro, A. E. (1984). Experiment and mathematics in Newton's theory of color. *Physics Today, 37*, 34–41.

Shapiro, W. A. (1966). Generalization of tristimulus coordinates. *Journal of the Optical Society of America, 56*, 795–802.

Sinden, R. H. (1923). Studies based on the spectral complementaries. *Journal of the Optical Society of America, 7*, 1123–1153.

Singer, H. W. (1906). Jakob Christoffel Le Blon and his three-colour prints. *Studio, 28*, 261–270.

Smith, T. (1931–1932). A reading of elementary colorimetric theory. *Transactions of the Optical Society (London), 33*, 214–227.

Smith, T., & Guild, J. (1931–1932). The C.I.E. colorimetric standards and their use. *Transactions of the Optical Society (London), 33*, 73–135.

Snyder, V., & Sisam, C. H. (1941). *Analytic geometry of space*. New York: Henry Holt.

Sperling, H. G. (1958). An experimental investigation of the relationship between colour mixture and luminous efficiency. In *Visual problems of colour* (pp. 321–325). Symposium at the National Physical Laboratory. London: Her Britannic Majesty's Stationery Office.

Stanlaw, J. M. (1987). *Color, culture, and contact: English loanwords and problems of color nomenclature in modern Japanese*. Unpublished doctoral dissertation, University of Illinois at Champaign-Urbana.

Stiles, W. S., & Burch, J. M. (1955). Interim report to the Commission Internationale de l'Éclairage, Zurich, on the National Physical Laboratory's investigation of colour-matching. *Optica Acta, 2*, 168–181.

Stiles, W. S., & Wyszecki, G. (1968). Intersections of the spectral reflectance curves of metameric object colors. *Journal of the Optical Society of America, 58*, 32–40.

Strang, G. (1988). *Linear algebra and its applications*. (3rd ed.). San Diego, CA: Academic Press.

Strutt, J. W. [Lord Rayleigh] (1889). The history of the doctrine of radiant energy. *Philosophical Magazine, 27*, 265–270.

Suppes, P., Krantz, D. M., Luce, R. D., & Tversky, A. (1989). *Foundations of measurement: Vol. 2: Geometrical, threshold, and probabilistic representations*. New York: Academic Press.

Takahama, K., & Nayatani, Y. (1972). New method for generating metameric stimuli of object colors. *Journal of the Optical Society of America, 62*, 1516–1520.

Tarrant, A. W. S. (1982). Light measurement. *Proceedings of the Society of Photo-Optical Instrumentation Engineers, 262.*

Thompson, B. [Benjamin, Count (of) Rumford] (1802). *Philosophical papers: Being a collection of memoirs, dissertations, and experimental investigations, together with letters to several persons on subjects connected with science and useful improvement.* London: Cadwell and Davies.

Thompson, B. [Benjamin, Count (of) Rumford] (1870–1875). *The complete works of Count Rumford.* (4 vols.). Boston: American Academy of Arts and Sciences.

Thompson, B. [Benjamin, Count (of) Rumford] (1968–1970). *Collected Works* (S. C. Brown, Ed.). (5 vols.). Cambridge, MA: Belknap Press of Harvard University Press.

Thompson, (Sir) William. [Lord Kelvin] (1852). On the mechanical action of radiant heat or light. *Philosophical Magazine, 4*, 256–260.

Thornton, W. A. (1972). Three-color visual response. *Journal of the Optical Society of America, 62*, 457–459.

Thornton, W. A. (1973). Matching lights, metamers, and human visual response. *Journal of Color and Appearance, 2*, 23–29.

Thornton, W. A (1974). Lamps for assessing metamerism. *Journal of the Illuminating Engineering Society, 4*, 11–18.

Thornton, W. A. (1978). Reply to Ohta-Wyszecki on location of nodes of metameric stimuli. *Color Research and Application, 3*, 202–204.

Thornton, W. A. (1986a). Evidence for the three spectral responses for the normal human visual system. *Color Research and Application, 11*, 160–163.

Thornton, W. A. (1986b). Note on visual responses: System vs. retinal. *Color Research and Application, 11*, 176–177.

Thornton, W. A. (1993a). Intersections of spectral power distributions that match. *Color Research and Application, 18*, 399–411.

Thornton, W. A. (1993b). Intersections of matching spectra: Applications. *Color Research and Application, 18*, 412–421.

Titchener, E. B. (1905). *A text-book of psychology.* New York: Macmillan.

Troland, W. L. T. (1922). Report of the Committee on Colorimetry for 1920–21. *Journal of the Optical Society of America and Review of Scientific Instruments, 6*, 537–596.

Trussell, H. J. (1991). Application of set theoretic systems to color systems. *Color Research and Application, 16*, 31–41.

Tyndall, E. P. T., & Gibson, K. S. (1924). Visibility of radiant energy equation. *Journal of the Optical Society of America, 9*, 403.

Vierordt, K. (1869). Beschreibung einer photometrischen Methode zur Messung und Vergleichung der Stärke des farbigen Lichtes. *Poggendorff's Annalen, 137*, 200–222. (Translated to English in *Annalen der physikalischen Chemie, 213*, 200–222.)

von Schelling, H. (1956). Concept of a distance in an affine geometry and its application in theories of vision. *Journal of the Optical Society of America, 46*, 309–315.

Wald, G. (1965). Frequency or wavelength? *Science, 150*, 1239–1240.

Walls, G. L. (1956). The G. Palmer story. *Journal of the History of Medicine and Allied Sciences, 11*, 66–96.

Walsh, J. W. T. (1925). Visibility of radiant energy equation. *Journal of the Optical Society of America, 11*, 111–112.

Walsh, J. W. T. (1963). *Photometry.* (3rd ed.). London: Constable.

Walters, H. V., & Wright, W. D. (1943). The spectral sensitivity of the fovea and extra fovea in the Purkinje range. *Proceedings of the Royal Society of London, 131B*, 340–361.

Warburton, F. L. (1936). The discrimination of the saturation of colours. *Proceedings of the Physical Society (London), 48*, 35–48.

Weale, R. A. (1957). Trichromatic ideas in the seventeenth and eighteenth centuries. *Nature (London), 179*, 648–651.

Wilson, M. H., & Brockelbank, R. W. (1955). Complementary hues of after-images. *Journal of the Optical Society of America, 45*, 293–299.

Wintringham, W. T. (1951). Color television and colorimetry. *Proceedings of the Institute of Radio Engineers, 39*, 1135–1172.

Wollaston, W. H. (1802). A method of examining reflective and dispersive powers, by prismatic reflection. *Transactions of the Royal Society of London, 92*, 365–388.

Wood, A. [completed by F. Oldham]. (1954). *Thomas Young: Natural philosopher, 1773–1829*. Cambridge: Cambridge University Press.

Woodworth, R. S. (1910). The puzzle of color vocabularies. *Psychological Bulletin, 7*, 325–334.

Wong, E. T. (1979). Generalized inverses as linear transformations. *Mathematical Gazette, 63*, 176–181.

Worthey, J. A. (1988). Calculation of metameric reflectances. *Color Research and Application, 13*, 76–84.

Wright, W. D. (1929). *A re-determination of the trichromatic color mixture data* (Special Rep. 139). London: Medical Research Council.

Wright, W. D. (1929–1930). A re-determination of the trichromatic coefficients of spectral colours. *Transactions of the Optical Society (London), 30*, 201–211.

Wright, W. D. (1949). *Photometry and the eye*. London: Hatton.

Wright, W. D. (1958). *The measurement of colour*. (2nd ed.). London: Adam Hilger Ltd. Original edition, 1944. 3rd ed., 1964. 4th ed., 1969. Distributed in the United States by Van Nostrand, New York.

Wünsch, C. E. (1792). *Versuche und Beobachtungen über die Farben des Lichts*. Leipzig: Breitkopf. (*Observations sur les couleurs de la lumière* appeared in *Annales de chimie, 64*, 135–163 [1807].)

Wyszecki, G. (1953). Valenzmetrische Untersuchung des Zusammenhanges zwischen normaler und anomaler Trichromasie. *Die Farbe, 2*, 39–52.

Wyszecki, G. (1958). Evaluation of metameric colors. *Journal of the Optical Society of America, 48*, 451–454.

Wyszecki, G. (1962). Metameric object colors. *Acta Chromatica, 1*, 1–6.

Wyszecki, G. (1967). Correlate for lightness in terms of CIE chromaticity coordinates and luminous reflectance. *Journal of the Optical Society of America, 57*, 254–257.

Wyszecki, G., & Sanders, C. L. (1957). Correlate for lightness in terms of CIE-tristimulus values. Part II. *Journal of the Optical Society of America, 47*, 840–842.

Wyszecki, G. & Stiles, W. S. (1982). *Color science: Concepts and methods, quantitative data and formulas*. (2nd ed.). New York: Wiley.

Wyszecki, G., & Wright, H. (1965). Field trial of the 1964 CIE color-difference formula. *Journal of the Optical Society of America, 55*, 1166–1174.

Yamauti, Z., & Okamatu, M. (1935). Interpolation of the standard visibility function to one millimicron steps. *Researches of the Electrotechnical Laboratory*, no. 388.

Young, T. (1802a). On the theory of light and colours. *Philosophical Transactions of the Royal Society of London, 92*, 12–48.

Young, T. (1802b). An account of some cases of the production of colours not hitherto described. *Philosophical Transactions of the Royal Society of London, 92*, 387–397.

Young, T. (1807). *A course of lectures on natural philosophy and the mechanical arts.* (2 vols.). London: J. Johnson.

Young, T. (1824). Chromatics. In supplement to the fourth, fifth, and sixth editions of the *Encyclopaedia Britannica*, with preliminary dissertations on the history of science. (6 vols.). Edinburgh: A. Constable.

Index

Typeset in 10/13 Minion
with Helvetica Neue Extended display
Designed by Paula Newcomb
Composed by Archetype
for the University of Illinois Press
Manufactured by Cushing-Malloy, Inc.

University of Illinois Press
1325 South Oak Street
Champaign, IL 61820-6903
www.press.uillinois.edu